# Springer Theses

Recognizing Outstanding Ph.D. Research

## Aims and Scope

The series "Springer Theses" brings together a selection of the very best Ph.D. theses from around the world and across the physical sciences. Nominated and endorsed by two recognized specialists, each published volume has been selected for its scientific excellence and the high impact of its contents for the pertinent field of research. For greater accessibility to non-specialists, the published versions include an extended introduction, as well as a foreword by the student's supervisor explaining the special relevance of the work for the field. As a whole, the series will provide a valuable resource both for newcomers to the research fields described, and for other scientists seeking detailed background information on special questions. Finally, it provides an accredited documentation of the valuable contributions made by today's younger generation of scientists.

## Theses are accepted into the series by invited nomination only and must fulfill all of the following criteria

- They must be written in good English.
- The topic should fall within the confines of Chemistry, Physics, Earth Sciences, Engineering and related interdisciplinary fields such as Materials, Nanoscience, Chemical Engineering, Complex Systems and Biophysics.
- The work reported in the thesis must represent a significant scientific advance.
- If the thesis includes previously published material, permission to reproduce this must be gained from the respective copyright holder.
- They must have been examined and passed during the 12 months prior to nomination.
- Each thesis should include a foreword by the supervisor outlining the significance of its content.
- The theses should have a clearly defined structure including an introduction accessible to scientists not expert in that particular field.

More information about this series at http://www.springer.com/series/8790

Adam Smith

# Disorder-Free Localization

Doctoral Thesis accepted by
the University of Cambridge, Cambridge, UK

*Author*
Dr. Adam Smith
Fakultät für Physik
Technische Universität München
Garching, Germany

*Supervisor*
Dmitry Kovrizhin
Clarendon Laboratory, Rudolf Peierls Centre
For Theoretical Physics
University of Oxford
Oxford, UK

ISSN 2190-5053 ISSN 2190-5061 (electronic)
Springer Theses
ISBN 978-3-030-20853-0 ISBN 978-3-030-20851-6 (eBook)
https://doi.org/10.1007/978-3-030-20851-6



This Springer imprint is published by the registered company Springer Nature Switzerland AG
The registered company address is: Gewerbestrasse 11, 6330 Cham, Switzerland

*To my parents.*

# Supervisor's Foreword

It is my pleasure to write this foreword for Adam Smith's thesis.

Since its discovery, the phenomenon of Anderson localization has been found to play an ever-increasing role in our understanding of condensed matter physics, from the quantum Hall effect to more recent studies of many-body localization.

This thesis presents another unexpected manifestation of Anderson localization phenomenology in the setting of $\mathbb{Z}_2$ lattice gauge theories. Theoretical studies of lattice gauge theories have been an important topic in theoretical physics, in both high-energy and condensed matter communities. In the latter, they arise as effective descriptions of strongly correlated systems, and the models discussed in this thesis are closely related to the famous Hubbard model and Kitaev's toric code.

This work may shed a new light on the physics of these models by looking at their behaviour in a non-equilibrium setting, where there is apparently a very rich phenomenology, which bridges between recent important ideas in quantum disentangled liquids and many-body localization.

One of the appeals of this work for me is that it shows conclusively for the first time that localization is possible in the absence of disorder. Such a possibility was first considered in early works on He-3 and He-4 mixtures in the 80s by my Ph.D. advisors Yuri Kagan and Leonid Maksimov. Since then, many people have been trying to devise models where disorder-free localization manifests itself. Adam's work presents such a setting for the first time.

From the technical perspective, this work is also quite beautiful. To my knowledge, this is the first time one can perform calculations essentially in a thermodynamic limit because of self-averaging property due to emergent disorder. This remarkable feature of the model also allows one to study out-of-time-ordered correlators, which have recently attracted a great deal of attention. For a microscopic quantum many-body system, this task has previously been achieved only in a very small number of systems, including the quantum Ising model and rather special quantum circuit models.

Using this opportunity, I would like to thank our collaborators Johannes Knolle and Roderich Moessner. It has been a pleasure to work with both of them over the years, and Adam is lucky to have been working closely with them during his Ph.D.

Non-equilibrium problems in strongly correlated systems, and more recently including lattice gauge theories, have been playing an increasingly important role in the studies of interacting many-body systems. Looking ahead, I hope the results of this thesis will find useful applications in these activities.

Oxford, UK
April 2019

Dmitry Kovrizhin

# Abstract

The venerable phenomena of Anderson localization, along with the more recent many-body localization (MBL), depend crucially on the presence of disorder. Here we introduce a family of simple translationally invariant models of fermions locally coupled to spins, which have a disorder-free mechanism for localization. This mechanism is due to a local $\mathbb{Z}_2$ gauge symmetry, and we uncover the connection to lattice gauge theories. We diagnose the localization through long-time memory of initial conditions after a global quantum quench.

One of the defining features of the models that we study is the binary nature of the emergent disorder, related to the $\mathbb{Z}_2$ degrees of freedom. This results in a qualitatively different behaviour in the strong effective disorder limit compared to typically studied models of localization. For example, it gives rise to the possibility of a delocalization transition via quantum percolation in higher than one dimension.

In connection to the recently proposed quantum disentangled liquid (QDL), we also study the entanglement properties of our models. The QDL provides an alternative to both complete localization and to the eigenstate thermalization hypothesis. Our models highlight the subtlety of defining a QDL, and we offer new insights into their entanglement properties.

While the simplest models we consider can be mapped onto free fermions, we also include interactions which lead to MBL-like behaviour characterized by logarithmic entanglement growth. We further consider interactions that generate dynamics for the conserved charges, which give rise to only transient localization behaviour, or quasi-MBL.

Finally, we present a proposal for the experimental measurement of gauge field correlators for our model in two dimensions. This proposal is based on interferometric techniques which are feasible using current experimental capabilities. Furthermore, the interacting generalizations of our models can be similarly implemented in experiments, providing access to the dynamics of strongly interacting lattice gauge theories, beyond what can be simulated on a classical computer.

## Publications Related to this Thesis

The core part of the thesis is based on research appearing in:

[1] A. Smith, J. Knolle, R. Moessner, and D. L. Kovrizhin, "*Logarithmic spreading of out-of-time-ordered correlators without many-body localization*", arXiv:1812.07981 (preprint)

[2] A. Smith, J. Knolle, R. Moessner, and D. L. Kovrizhin, "*Dynamical localization in $\mathbb{Z}_2$ lattice gauge theories*", Phys. Rev. B 97, 245137 (2018)

[3] A. Smith, D. L. Kovrizhin, R. Moessner, and J. Knolle, "*Dynamics of a lattice gauge theory with fermionic matter – minimal quantum simulator with time-dependent impurities in ultracold gases*", Quantum Sci. Technol. 3, 044003 (2018)

[4] A. Smith, J. Knolle, R. Moessner, and D. L. Kovrizhin, "*Absence of Ergodicity without Quenched Disorder: from Quantum Disentangled Liquids to Many-Body Localization*", Phys. Rev. Lett. 119, 176601 (2017)

[5] A. Smith, J. Knolle, D. L. Kovrizhin, and R. Moessner, "*Disorder-free localization*", Phys. Rev. Lett. 118, 266601 (2017)

Other works published during my doctoral research not included in this thesis:

[6] A. Smith, J. Knolle, D. L. Kovrizhin, J. T. Chalker, and R. Moessner, "*Majorana spectroscopy of three-dimensional Kitaev spin liquids*", Phys. Rev. B 93, 235146 (2016)

[7] A. Smith, J. Knolle, D. L. Kovrizhin, J. T. Chalker, and R. Moessner, "*Neutron scattering signatures of the 3D hyperhoneycomb Kitaev quantum spin liquid*", Phys. Rev. B 92, 180408(R) (2015)

# Acknowledgements

First and foremost, I thank Dima for his invaluable support and guidance. He introduced me to condensed matter physics and has shaped the physicist that I am today, and for that I am truly grateful. He has given me the freedom to pursue whatever I find interesting and has helped me to push myself to achieve more than I thought I could.

I would like to give special thanks to Johannes, who has been a mentor and a role model for me during my time in Cambridge. He has always been available for advice, whether it be about physics or important life decisions.

During my Ph.D., I have had the privilege to collaborate with Roderich Moessner and John Chalker. Both of these men are truly inspirational. I thank them for our collaborations, but also for their advice on life in academia.

Thanks to Claudio Castelnovo, Fabian Essler, Adam Nahum, Thomas Veness and Ulrich Schneider for enlightening discussions with myself and my co-authors during the writing of the publications that form the core of this thesis.

I am grateful to the condensed matter theory groups in Nottingham, Cologne, and at the TU Munich, for inviting me to present my work, for the stimulating discussions and most of all for their hospitality.

A huge thank you to everyone in TCM who has made the last 3 years so enjoyable. I would particularly like to thank Dima Khmelnitskii for endeavouring educate the TCM Ph.D. students and for answering my many questions. The Ph.D. students and post-docs in TCM are too numerous to list here, but I would like to thank them all. I am indebted to Daniel, Ben, Eze, Philip, Petr, Ollie and Stephen for proofreading chapters of my thesis.

I would especially like to thank my officemates and friends Daniel, Beñat and Chris. I would not understand half the physics that I do without our discussions. We won't loose touch but I will truly miss sharing an office with you all. Cheers! (not chairs).

I thank my incredible girlfriend Sarah, for putting up with me, and always reminding me that there is much more to life than physics.

Finally, I am thankful for my family. I can always rely on their love and support. I would have dropped maths and physics after high-school were it not for my Mum and Dad—look at me now!

# Contents

# Chapter 1
# Introduction

Nature is extremely complex! Not necessarily because of the fundamental laws that govern it, but due to the unfathomable number of degrees of freedom interacting with each other. Fortunately, many natural phenomena display a remarkable level of universality, that is to say, they are independent of their microscopic details. However, there are many situations beyond this universal behaviour that are particularly interesting. Perhaps the most important are non-equilibrium phenomena, which cover the vast majority of what we observe in our everyday lives—car engines, the weather, and life itself are but a few examples. In this thesis we are interested in questions about the relaxation of isolated quantum systems after they are taken far from equilibrium.

The observed universality of certain phenomena in nature is encapsulated in statistical mechanics. Rather than considering the full classical or quantum mechanical evolution of a many-particle system, the idea is instead to consider many copies of the system—a so-called statistical ensemble. The problem is then reduced to studying probability distributions with respect to this ensemble [1–4]. This allows us to describe the (close to[1]) equilibrium behaviour of a system using only a handful of physical parameters such as temperature, pressure and the number of particles. Take for example the air around us which is composed of several different molecular gases and water vapour and consists of $\sim 10^{25}$ molecules per cubic square meter. Nevertheless, for most purposes it can be described in terms of its temperature, pressure, or humidity, and we rarely have to think about the behaviour of individual molecules.

In this simpler description provided by statistical mechanics it is possible to understand ordered phases of matter and the transitions between them. Taking these ideas a step further, Lev Landau [8] established a phenomenological theory that takes as its starting point a quantity called an order parameter that distinguishes the phases, and the symmetries of the system in different states [8]. Despite this reduced description,

[1]There are well developed tools such as linear response theory and the fluctuation-dissipation theorem that allow non-equilibrium behaviour, such as thermal and electrical conductivity, to be understood from an equilibrium theory [5–7].



A. Smith, *Disorder-Free Localization*, Springer Theses,
https://doi.org/10.1007/978-3-030-20851-6_1

the resulting concept of symmetry breaking has proven extremely powerful in explaining phases and phase transitions in condensed matter systems [7, 9].

Away from equilibrium we no longer have a single unifying framework analogous to statistical mechanics. While there exist formalisms such as kinetic theory [10] or Keldysh field theory [11, 12], these are either incomplete or are in most cases technically challenging. Typically, dynamical problems have to be tackled on a case-by-case basis using a range of physical approximations. Often one must resort to numerical simulations such as exact diagonalization methods which are limited to approximately 20 spins or fermions.

Recently there has been renewed interest in how a closed quantum mechanical system can relax and the nature of the possible long-time equilibrium behaviour. Classical systems are able to thermalize at long times—that is, all parts of the system will be in thermal equilibrium with each other—due to non-linear and chaotic behaviour. However, quantum mechanical systems obey unitary time evolution and thus in this sense cannot be chaotic [13]. It is therefore a natural question to ask if and how a quantum system can relax, and how this relates to thermalization behaviour observed in nature. An answer to the latter question was suggested in the form of the eigenstate thermalization hypothesis, which suggests that eigenstates of thermalizing quantum systems themselves locally look thermal [14–16], a statement that we will make more precise in the following sections.

The resurgence of interest in quantum relaxation comes in part from a violation of the eigenstate thermalization hypothesis in the dynamical many-body localized phase of matter [17, 18]. Such systems avoid thermalization and for instance can retain memory of initial states indefinitely, due to the presence of disorder. These systems are important not only because they challenge our theoretical understanding of complex quantum systems but they may also find application in future technology. For example, this protection of information in the initial state could potentially be used in quantum memory devices and in quantum computations.

## 1.1 Thesis Outline

In this thesis we contribute to answering questions about relaxation of isolated quantum systems. In particular, we provide the answer to a long-standing question about the role of disorder in localized systems—we demonstrate that disorder is not a prerequisite condition for localization. Further, we identify a connection between our disorder-free mechanism for localization and lattice gauge theories. The latter are themselves of fundamental importance in the description of strongly correlated phases of matter [19–21] and we make connections with a number of theoretical models.

In this chapter we introduce the main concepts that will be used throughout this thesis. We first define the protocol for taking our system far from equilibrium and discuss the subsequent spreading of correlations and the long-time behaviour. Here we present the eigenstate thermalization hypothesis and the formalism of density matrices and entanglement, which we will refer to throughout. In Sect. 1.3 we give a brief overview of the localization phenomena which provides an alternative scenario to thermalization. Following this, we give a brief history of the question we are concerned with in this thesis—namely, whether localization is possible without disorder. In Sect. 1.5 we introduce gauge theories with some examples. The final part of the introduction concerns the progress in experiment, relevant to both localization physics and to the physics of lattice gauge theories.

The remainder of the thesis is structured as follows. In Chap. 2 we introduce our model. We provide details of a mapping to free fermions and identify conserved charges, which is central to our ability to study large systems and understand the disorder-free mechanism for localization. Chapter 3 constitutes a numerical investigation of the localization in our model. We also make a connection to a quantum percolation problem not found in models of localization with continuously sampled disorder. The notion of a quantum disentangled liquid is discussed in Chap. 4. We reveal the subtleties in its definition and offer more insights into the entanglement properties of our model. In Chap. 5 we investigate the out-of-time-ordered correlators. These quantify operator spreading and we reveal a wide range of behaviour. We discuss possible extensions to our model in Chap. 6 that take it away from the free fermion limit by adding interactions. Finally, in Chap. 7 we propose experimental protocols in cold atoms for simulating dynamics of the gauge field in our model and for the measurement of correlators. We close the thesis with a discussion of our results, open questions, and directions for future research.

## 1.2 Quantum Quenches and Thermalization

In this thesis we are concerned with quantum quenches in closed systems and investigate the properties of quantum models using protocols that take the system far from equilibrium. A quantum quench is an instantaneous change in the Hamiltonian of the system [22, 23]. The system is initially prepared in some state $|\psi(0)\rangle$, which we will consider to be the ground state of a preparation Hamiltonian $\hat{H}_0$ at time $t = 0$. The Hamiltonian is then instantaneously quenched at $t = 0$ to $\hat{H}$ and the wavefunction evolves as

$$|\psi(t)\rangle = e^{-i\hat{H}t}|\psi(0)\rangle. \tag{1.1}$$

For our purposes we can prepare the state $|\psi(0)\rangle$ by hand but in experiments the systems will be prepared close to the ground state of a Hamiltonian $\hat{H}_0$ and ramped as quickly as possible, faster than any time scale set by $\hat{H}$, to approximate the true instantaneous quench.

A quantum quench can either be local or global, depending on how the Hamiltonian $\hat{H}$ differs from the preparation Hamiltonian $\hat{H}_0$, but in this thesis we will consider only global quenches. Following a quench there are then two natural things to study—the spreading of correlations and the properties of the system at long times.

While we only consider global quantum quenches, we will also encounter local quenches in the calculation of certain correlators. Local quenches also have an important place in condensed matter physics, most notably in the X-ray edge problem [24]. There also exists an exact mapping of the dynamical structure factor of Kitaev quantum spin liquids [25] to the X-ray edge problem [26]. Based on the X-ray edge solution, recently an exact method was developed to calculate the full inelastic neutron scattering response for the honeycomb Kitaev model [27], which has since been compared with experiments on a candidate material [28]. During my PhD I also contributed to this field by developing an exact method for three-dimensional Kitaev models and studied qualitative behaviour of dynamical correlation functions in three dimensions [29, 30].

### 1.2.1 Spreading of Correlations

Let us first consider time-dependent correlations after a quantum quench. If the Hamiltonian is local, as will always be the case for us, immediately after the quench the information of the change in the Hamiltonian is only known locally. As the system evolves, correlations build up over larger distances. Let us consider, for concreteness, the example of free fermions described by a 1D tight-binding Hamiltonian

$$\hat{H} = -J \sum_j \left( \hat{c}_j^\dagger \hat{c}_{j+1} + \text{H.c.} \right), \tag{1.2}$$

with canonical anti-commutation relations $\{\hat{c}_i^\dagger, \hat{c}_j\} = \delta_{ij}$ and $J$ constant. We are interested in the correlator of particle densities

$$\langle \psi(t) | \hat{n}_j \hat{n}_k | \psi(t) \rangle_c = \langle \psi(t) | \hat{n}_j \hat{n}_k | \psi(t) \rangle - \langle \psi(t) | \hat{n}_j | \psi(t) \rangle \langle \psi(t) | \hat{n}_k | \psi(t) \rangle. \tag{1.3}$$

Considering the initial state $|\psi(0)\rangle$ to be a charge density wave with the odd sites occupied and even sites empty, i.e., $|\psi(0)\rangle = |\cdots 101010 \cdots\rangle$, we impose periodic boundary conditions and take the thermodynamic limit, where we can compute the correlator exactly. The analytic result is

$$\langle \psi(t) | \hat{n}_j \hat{n}_k | \psi(t) \rangle_c = \frac{1}{4} \delta_{j,k} - \frac{1}{4} J_{|j-k|}(4Jt)^2, \tag{1.4}$$

where $J_n(x)$ are the Bessel functions of the first kind, see Appendix A for the calculation. The absolute value of this correlator is shown in Fig. 1.1a.

This correlator shows a clear linear light-cone which spreads with velocity $4J$, indicated in Fig. 1.1a, which is consistent with the bound on correlation spreading found by Lieb and Robinson [31]. This bound is on the spreading of operators whose important consequence [32] for our purposes is that

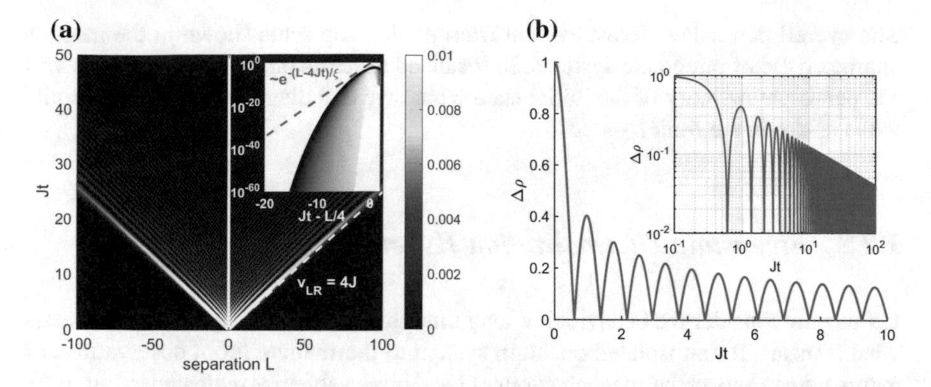


**Fig. 1.1** **a** Spreading of correlations for free fermions. Plotted is the absolute value of the connected density correlator $|\langle\psi(t)|\,\hat{n}_j\hat{n}_{j+L}\,|\psi(t)\rangle_c|$. The Lieb-Robinson velocity which leads to a linear light-cone is indicated by a dashed white line. (inset) Behaviour close to the light-cone on a log scale. (Time dependence for fixed separation $L$ is shifted by the light-cone velocity). **b** The decay of the density imbalance $\Delta\rho(t)$, defined in Eq. (1.6), which demonstrates the relaxation of an initial charge density wave. (inset) Same decay for longer times on a log-log scale

$$|\langle\psi(t)|\,\hat{O}_A\hat{O}_B\,|\psi(t)\rangle_c| \le c\,e^{-\frac{L-2vt}{\xi}}, \quad (1.5)$$

where $L$ is the spatial separation between the local operators $\hat{O}_A$ and $\hat{O}_B$, and $c, v, \xi$, are the constant prefactor, speed and length scale, which are to be determined.[2] In our case, the velocity is observed to be the maximal group velocity $v = 2J$ and we have exponentially small correlations outside the light-cone with Lieb-Robinson velocity $v_{\mathrm{LR}} = 4J$, as shown in the inset of Fig. 1.1a. This bound need not be saturated, particularly when we do not have long-lived quasi-particles, and we will later use deviation from this linear light-cone behaviour to diagnose localization.

Since we start with a charge density wave and evolve with a translationally invariant Hamiltonian, we expect that this inhomogeneous pattern should relax as a function of time. We can quantify the presence of the charge density wave using the density imbalance

$$\Delta\rho(t) = \frac{1}{N}\sum_n |\langle\psi(t)|\hat{n}_{n+1} - \hat{n}_n|\psi(t)\rangle|, \quad (1.6)$$

which takes the maximal value of 1 for a charge density wave and is zero for a translationally invariant state. We are also able to calculate this value exactly in the thermodynamic limit,

$$\Delta\rho(t) = |J_0(4Jt)|, \quad (1.7)$$

which is shown in Fig. 1.1b, see Appendix A for the calculation. This shows that the initial CDW does indeed relax and the system loses memory of the initial state.

[2] Note that this bound on correlators only applies for quenches from states with finite correlation length [32].

The overall power-law decay, evident from the log-log scale shown in the inset, is characteristic of integrable systems. In localized systems on the other hand we will see persistent memory of the initial state which we can diagnose by having a finite value of $\Delta\rho$ in the limit $t \to \infty$.

### 1.2.2 Eigenstate Thermalization Hypothesis

Let us now consider the behaviour at long times after a quantum quench and discuss what it means for an isolated quantum system to thermalize. From observations of nature we can expect thermal behaviour of local observables, as indicated in Fig. 1.2a. That is, while the short time behaviour of local observables may be non-universal and depend on the initial state, the long-time behaviour should be determined only by a few quantities such as the energy or effective temperature of the initial state. The eigenstate thermalization hypothesis (ETH) provides a possible answer to the question of how a closed quantum system can display this expected thermal behaviour. It says that in a thermalizing system the eigenstates themselves look thermal [14, 15]. Let us now make this statement more precise. In our exposition we will follow closely Ref. [18].

Let us consider a physical observable associated with the local operator $\hat{O}$, then the expectation value $\langle\psi(t)|\hat{O}|\psi(t)\rangle$ can be expanded in terms of energy eigenvalues and eigenvectors as

$$\langle\psi(t)|\hat{O}|\psi(t)\rangle = \sum_{\alpha\beta} C_\alpha^* C_\beta e^{i(E_\alpha - E_\beta)t} O_{\alpha\beta}, \tag{1.8}$$

where $O_{\alpha\beta} = \langle\alpha|\hat{O}|\beta\rangle$ are the matrix elements of the operator in the basis of energy eigenstates. The coefficients $C_\alpha$ come from the decomposition of the initial state in terms of eigenstates $|\psi(0)\rangle = \sum_\alpha C_\alpha|\alpha\rangle$. The eigenstate thermalization hypothesis then suggests the form of these matrix elements for a general local operators $\hat{O}$,

$$O_{\alpha\beta} = O(E)\delta_{\alpha\beta} + e^{-S(E)/2} f(E, \omega) R_{\alpha\beta}, \tag{1.9}$$

where $E = (E_\alpha + E_\beta)/2$ is the average energy, $\omega = E_\alpha - E_\beta$ is the energy difference, $R_{\alpha\beta}$ are random numbers with zero mean and unit variance, and $f(E, \omega)$ is a smooth function of $E$ assumed to decay with $|\omega|$. In the exponential there appears the thermodynamic entropy $S(E) = -\mathrm{Tr}(\rho \log \rho)$, where $\rho = e^{-\beta\hat{H}}/\mathcal{Z}$ with partition function $\mathcal{Z} = \mathrm{Tr}[e^{-\beta\hat{H}}]$ and the effective temperature $\beta$ defined via $E = \mathrm{Tr}[e^{-\beta\hat{H}}\hat{H}]/\mathcal{Z}$. Equation (1.9) says that the eigenstates are effectively random vectors in the Hilbert space but that on average they locally look thermal in the sense that the diagonal elements $O_{\alpha\alpha}$ are given by their thermal expectation value $O(E) = \mathrm{Tr}[e^{-\beta\hat{H}}\hat{O}]/\mathcal{Z}$.

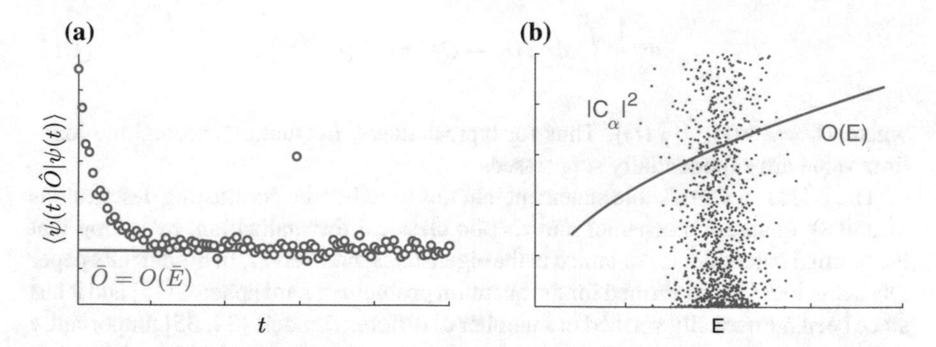


**Fig. 1.2** **a** Expected behaviour of local observables in a thermalizing quantum system. The long-time value is determined by the energy of the initial state but short-time behaviour is non-universal. There are fluctuations around this long-time value which can be large on rare occasion. **b** Illustration of the realistic constraints on smoothness of the diagonal elements of local operators and the narrow energy distribution for typical initial states. Figures are inspired by the KITP lecture Ref. [33]

From the ansatz for the matrix elements in Eq. (1.9) we can then find the relaxation of observables after a quantum quench. Plugging Eq. (1.9) into the time evolution (1.8) we see that the function $f(E, \omega)$ determines the non-universal corrections to the time evolution of the observable, which in the long-time limit are washed away due to rapidly oscillating phase factors. What is left are the diagonal elements, which are given by thermal averages $O(E) = \mathrm{Tr}[e^{-\beta\hat{H}}\hat{O}]/\mathcal{Z}$. The conclusion of ETH is that the infinite-time average of local observables,

$$\bar{O} = \lim_{t\to\infty}\frac{1}{t}\int_0^t \mathrm{d}\tau\ \langle\psi(\tau)|\hat{O}|\psi(\tau)\rangle = \sum_\alpha |C_\alpha|^2 O(E_\alpha) \approx O(\bar{E}), \tag{1.10}$$

is independent of the details of the initial state $|\psi(0)\rangle$ and depends only on the expectation value of the energy $\bar{E} = \langle\psi(0)|\hat{H}|\psi(0)\rangle$. The final approximate equality assumes a relatively weak restriction on the coefficients $|C_\alpha|^2$, namely that they are appreciable only over a small energy window, but otherwise unrestricted. This energy window should be small enough that the function $O(E)$ varies at most linearly over this window,[3] see Fig. 1.2b The result is that the infinite time average of the operator is thermal, i.e. $\bar{O} \approx \mathrm{Tr}[e^{-\beta\hat{H}}\hat{O}]/\mathcal{Z}$ which depends only on the effective temperature of the initial state and not, for example, on any spatial structure. Note that without the extra conditions on the smoothness of $O(E)$ and the small energy window for the coefficients, the effective energy/temperature would depend on the observable, which is not expected for a generic local observable.

Note that the hypothesised form for the matrix elements of the local observables in Eq. (1.9) also contains information about the fluctuations around the infinite time average value. The average fluctuations are given by

[3] In the original paper by Srednicki [15] it was assumed that $O(E)$ is approximately constant over this energy window, although this seems to be stricter than necessary.

$$\lim_{t\to\infty}\frac{1}{t}\int_0^t \mathrm{d}\tau\ (O_\tau-\bar{O})^2=\mathcal{O}\left(e^{-S(E)}\right), \tag{1.11}$$

where $O_t=\langle\psi(t)|\hat{O}|\psi(t)\rangle$. Thus for typical states, fluctuations around the long-time value are exponentially suppressed.

The ETH is a remarkable statement that ties together the contrasting descriptions of unitary quantum mechanics and chaotic classical thermalization, by stating that the thermal behaviour is contained in the eigenstates themselves. In Srednicki's paper this behaviour was confirmed for the quantum problem of hard spheres [15] and it has since been numerically verified in a number of different models [34, 35]. Importantly though, many-body localization provides a robust setting where ETH is violated. It can be characterized, for example, by the persistent memory of initial states such as a charge density wave.

### 1.2.3 Density Matrices and Entanglement

An alternative and natural way to define thermalization is in terms of density matrices [1, 2]. The density matrix arises as a way to describe both pure quantum states and statistical mixtures in the same formalism. Within this formalism it is then possible to formulate measures of the quantum entanglement, which roughly speaking is the dependence of the state of a subsystem on the rest of the system. The density matrix formalism, as well as the concept of quantum entanglement [36, 37], will be used throughout this thesis and we therefore take a short detour to introduce these concepts.

The density matrix is a description of the state of a quantum system which can be in either a pure quantum state or a mixed state that is a statistical mixture of pure states. If our system is in the pure state $|\psi\rangle$, then the density matrix is defined as $\rho=|\psi\rangle\langle\psi|$, i.e., a projector onto this state. More generally, the density matrix is given by a statistical mixture $\rho=\sum_\alpha\lambda_\alpha|\alpha\rangle\langle\alpha|$, where $\{|\alpha\rangle\}$ is some set of pure states. If we are in a thermal state with inverse temperature $\beta$, for example, then the density matrix is given by

$$\rho=\frac{1}{\mathcal{Z}}e^{-\beta\hat{H}}=\sum_\alpha\frac{e^{-\beta E_\alpha}}{\mathcal{Z}}|\alpha\rangle\langle\alpha|, \tag{1.12}$$

where $|\alpha\rangle$ are eigenstates of the Hamiltonian $\hat{H}$ with energy $E_\alpha$, and $\mathcal{Z}=\mathrm{Tr}[e^{-\beta\hat{H}}]$ is the partition function which normalizes the density matrix. The density matrix is normalized such that the total probability, $\mathrm{Tr}[\rho]=1$. We also have that $\mathrm{Tr}[\rho^2]\leq 1$ with equality if and only if $\rho$ describes a pure state, because it is a projector.[4]

[4]Proof: If $\rho$ is a projector then $\rho^2=\rho$, and $\mathrm{Tr}[\rho^2]=\mathrm{Tr}[\rho]=1$. On the other hand if $\mathrm{Tr}[\rho^2]=\mathrm{Tr}[\rho]=1$, then in terms of the eigenvalues $1\geq\lambda_\alpha\geq 0$ of $\rho$ we have $\sum_\alpha\lambda_\alpha^2=\sum_\alpha\lambda_\alpha=1$. Suppose we have more than one non-zero eigenvalue, then we have $\lambda_\alpha<1,\forall\alpha$, and so $\sum_\alpha\lambda_\alpha^2<\sum_\alpha\lambda_\alpha$,

While there are many measures of entanglement used in the context of localization—and many more in quantum information theory—we choose to focus here on the von Neumann bipartite entanglement entropy. The most common alternatives are the family of Renyi entropies, which contain the von Neumann entropy as a limit.

We start with the density matrix of a pure state $\rho = |\psi\rangle\langle\psi|$. To define the bipartite entanglement entropy we must first split the system into two parts, subsystem $A$ and its complement $B$. It is then these subsystems that are said to be entangled. We proceed by defining the reduced density matrix on $A$ by tracing over states for $B$, i.e., $\rho_A = \mathrm{Tr}_B[\rho]$. If the subsystems $A$ and $B$ are entangled then this reduced density matrix must necessarily have a form of a mixed state and the entanglement entropy quantifies the mixing. The von Neumann entanglement entropy generalises the Gibbs entropy and is defined as

$$S = -\mathrm{Tr}[\rho_A \ln \rho_A] = -\sum_i \lambda_i \ln \lambda_i, \tag{1.13}$$

where $\lambda_i > 0$ are the eigenvalues of the reduced density matrix $\rho_A$.

To aid this definition let us look closer at the structure of the reduced density matrix and by considering an example. Any pure state can be written in terms of a basis of states $\{|i\rangle_A \otimes |j\rangle_B\}$ where $\{|i\rangle_A\}$ and $\{|i\rangle_B\}$ are orthonormal bases for the subsystems $A$ and $B$ respectively. In this basis the pure state can be written as

$$|\psi\rangle = \sum_{ij} \psi_{ij} |i\rangle_A \otimes |j\rangle_B. \tag{1.14}$$

It is then possible to perform singular value decomposition of the rectangular matrix with elements $\psi_{ij}$. This gives $\psi_{ij} = \sum_k U_{A,ik}\sqrt{\lambda_k} U_{B,kj}$, where $\lambda_k > 0$. By redefining the basis states for the two subsystems as $|\tilde{i}\rangle_A = \sum_j U^T_{A,ij} |i\rangle_A$ and $|\tilde{i}\rangle_B = \sum_j U_{B,ij} |i\rangle_b$, the state can be written as

$$|\psi\rangle = \sum_k \sqrt{\lambda_k} |\tilde{k}\rangle_A \otimes |\tilde{k}\rangle_B, \tag{1.15}$$

and the reduced density matrix becomes

$$\rho_A = \sum_k \lambda_k |\tilde{k}\rangle_A \langle\tilde{k}|_A, \tag{1.16}$$

with $\lambda_k > 0$ and $\sum_k \lambda_k = 1$. From the decomposition of the state in Eq. (1.15) we can see that the reduced density matrix, $\rho_B$, for subsystem $B$, i.e., tracing out subsystem $A$, has the same eigenvalues as $\rho_A$. The von Neumann entropy is then independent of which part of the system one traces out.

---

which is a contradiction. Thus we have one non-zero eigenvalue, which is equal to 1, i.e., $\rho$ is a projector.

**Example**

Let us study an example which will be instructive in clarifying the above definitions of density matrices and entanglement. We will consider a 4 state system composed of two spin 1/2 degrees of freedom, where we define subsystems $A$ and $B$ as a single spin each. We will consider the system in one of two states: $|\psi_{\mathrm{TP}}\rangle = \frac{1}{2}(|\uparrow\rangle_A + |\downarrow\rangle_A) \otimes (|\uparrow\rangle_B + |\downarrow\rangle_B)$, which is a tensor product of states on systems $A$ and $B$; and $|\psi_{\mathrm{EPR}}\rangle = \frac{1}{\sqrt{2}}(|\uparrow\rangle_A|\uparrow\rangle_B + |\downarrow\rangle_A|\downarrow\rangle_B)$, the Einstein-Podolsky-Rosen (EPR) pair or Bell pair state, which cannot be decomposed as a tensor product of states on $A$ and $B$ separately.

Let us then consider the basis $\{|\uparrow\uparrow\rangle, |\uparrow\downarrow\rangle, |\downarrow\uparrow\rangle, |\downarrow\downarrow\rangle\}$ for the combined 4 state system. In this basis the density matrices can be written as

$$\rho_{\mathrm{TP}} = \begin{pmatrix} \frac{1}{4} & \frac{1}{4} & \frac{1}{4} & \frac{1}{4} \\ \frac{1}{4} & \frac{1}{4} & \frac{1}{4} & \frac{1}{4} \\ \frac{1}{4} & \frac{1}{4} & \frac{1}{4} & \frac{1}{4} \\ \frac{1}{4} & \frac{1}{4} & \frac{1}{4} & \frac{1}{4} \end{pmatrix} \quad \rho_{\mathrm{EPR}} = \begin{pmatrix} \frac{1}{2} & 0 & 0 & \frac{1}{2} \\ 0 & 0 & 0 & 0 \\ 0 & 0 & 0 & 0 \\ \frac{1}{2} & 0 & 0 & \frac{1}{2} \end{pmatrix}. \tag{1.17}$$

The reduced density matrices for subsystem $A$ (equivalently $B$) using the basis $\{|\uparrow\rangle, |\downarrow\rangle\}$ are then given by

$$\rho_{\mathrm{TP},A} = \begin{pmatrix} \frac{1}{2} & \frac{1}{2} \\ \frac{1}{2} & \frac{1}{2} \end{pmatrix} \quad \rho_{\mathrm{EPR},A} = \begin{pmatrix} \frac{1}{2} & 0 \\ 0 & \frac{1}{2} \end{pmatrix}. \tag{1.18}$$

The reduced density matrix $\rho_{\mathrm{TP},A}$ is equivalent to the full density matrix of a single spin 1/2 in the pure state $(|\uparrow\rangle + |\downarrow\rangle)/\sqrt{2}$ and thus the existence of subsystem $B$ is of no relevance for measurements of subsystem $A$. For the EPR pair, however, $\rho_{\mathrm{EPR},A}$ cannot be written as the density matrix of a pure state. The ignorance of subsystem $B$ means that subsystem $A$ is in an effective statistical combination of the two states and there is no measurement that can be performed on that subsystem alone that has a determined outcome. The corresponding entanglement entropies are $S_{\mathrm{TP}} = 0$ and $S_{\mathrm{EPR}} = \ln 2$. For the EPR pair the subsystem $A$ is 'entangled' with subsystem $B$, and in fact is a maximally entangled quantum state.[5] Note that the spectra of $\rho_{\mathrm{TR},A}$ and $\rho_{\mathrm{EPR},A}$ are $\{1, 0\}$ and $\{\frac{1}{2}, \frac{1}{2}\}$, respectively.

The von Neumann entanglement entropy takes a matrix and produces a single number. The reduced density matrices therefore contain far more information about the entanglement in the system that the entanglement entropy. More information can be extracted by looking at the spectrum of the reduced density matrix $\lambda_i = e^{-\chi_i}$, where the eigenvalues $\chi_i$ are referred to as the entanglement spectrum [38, 39].

---

[5] Up to unitary transformations, the EPR pair is the unique maximally entangled state for a 2 qubit system.

### 1.2.4 Thermalization for Density Matrices

Equipped with density matrices we can provide an alternative definition of thermalization. It is believed that in many situations this definition is equivalent to the ETH [22], although we will not discuss this here. For the purposes of this thesis we can take any one as our definition of thermalization.

Let us consider the density matrix after the quantum quench $\rho(t) = |\psi(t)\rangle\langle\psi(t)|$ and let us take a partition $A$ of the system, with $B$ the complement of $A$. The system thermalizes if the limit of the density matrix for $t \to \infty$ exists and is equal to

$$\lim_{t\to\infty} \mathrm{Tr}_B \left[|\psi(t)\rangle\langle\psi(t)|\right] = \mathcal{Z}^{-1}\mathrm{Tr}_B \left[e^{-\beta\hat{H}}\right] = \mathrm{Tr}_B \left[|\alpha\rangle\langle\alpha|\right], \tag{1.19}$$

with the inverse temperature $\beta$ defined by $\bar{E} = \mathcal{Z}^{-1}\mathrm{Tr}_B[e^{-\beta\hat{H}}\hat{H}]$, where $\bar{E} = \langle\psi(t)|\hat{H}|\psi(t)\rangle$ is the energy of the state, and $|\alpha\rangle$ is an eigenstate of $\hat{H}$ with $E_\alpha = \bar{E}$. Strictly speaking we should take the limit $|B| \to \infty$, with $|A|$ kept fixed, before taking the limit $t \to \infty$, where $|\cdot|$ denotes the size of the partition. The first equality says that the system thermalizes if the density matrix at long times locally looks like a thermal density matrix for the Gibbs ensemble. The second equality says that this in turn is locally equivalent to a density matrix for the microcanonical ensemble. The latter is essentially the statement of eigenstate thermalization—that the eigenstates themselves locally look thermal.

For us, the most important consequence of thermalization is the long-time behaviour of observables (in the thermodynamic limit $|B| \to \infty$, with $|A|$ fixed), where we find

$$\lim_{t\to\infty} \mathrm{Tr}_B \left[\rho(t)\hat{O}\right] = \mathcal{Z}^{-1}\mathrm{Tr}_B \left[e^{-\beta\hat{H}}\hat{O}\right] = O_{\alpha\alpha}, \tag{1.20}$$

which matches the conclusion of the ETH as discussed in Sect. 1.2.2, with the same effective temperature $\beta$ defined by $\mathrm{Tr}_B[\rho(t)\hat{H}] = \mathcal{Z}^{-1}\mathrm{Tr}_B[e^{-\beta\hat{H}}\hat{H}]$.

## 1.3 Localization

We now turn our attention to localization in quantum systems, which provides an alternative scenario to thermalization and violates the ETH. Localization was first described by P. W. Anderson in 1958 in the seminal paper [40] on the "Absence of diffusion in certain random lattices". It is a crucial ingredient to understanding the resistive properties of metals [5, 6, 41], magnetoresistence [42–44], and also in the explanation of quantised plateaux in quantum Hall systems [5, 41, 45].

Anderson's original goal was to understand the effects of disorder on interacting quantum systems, but was restricted to the study of the single-particle phenomenon in Ref. [40]. It was not until recently that progress has been made on the interacting

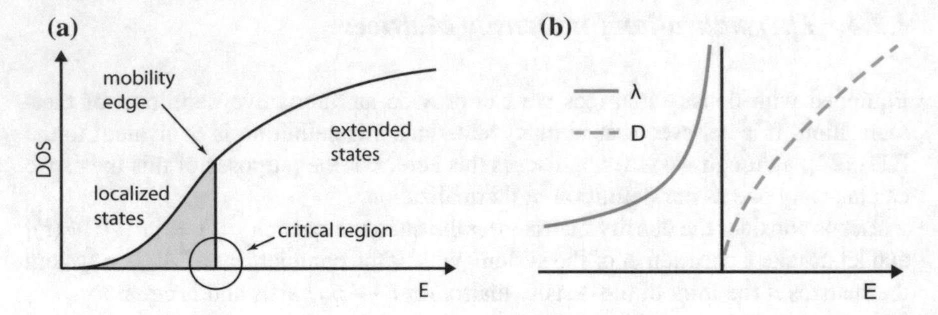


**Fig. 1.3** **a** Schematic picture of the density of states near a band tail in the presence of disorder (in 3D). The states in the tail are localized and separated from the extended states by a mobility edge. **b** Behaviour of the localization length $\lambda$ and diffusion coefficient $D$ in the critical region close to the mobility edge

case, dubbed many-body localization (MBL) [46–48]. Since then there have been many advances in our understanding of the properties of MBL, both analytic and numerical, but there still remains many open questions [18].

## 1.3.1 Anderson Localization

Let us begin by discussing the single-particle physics of Anderson localization. To describe its phenomenology, we consider a tight-binding Hamiltonian for spinless fermions with disorder potential

$$\hat{H} = -J \sum_{\langle jk \rangle} \hat{c}_j^\dagger \hat{c}_k + \sum_j V_j \hat{c}_j^\dagger \hat{c}_j, \tag{1.21}$$

where $\langle jk \rangle$ indicates nearest neighbour sites, and $V_j$ uniformly sampled from the interval $[-W, W]$. This was the original model discussed by Anderson and remains the typical model of Anderson localization, along with its continuum counterpart.

In three dimensions, the presence of disorder in the potential leads to tails at the band edge with localized states separated by a mobility edge from extended states in the centre of the band [5, 49], see Fig. 1.3a. The eigenfunctions in the tails are localized in space with an exponential envelope, $\psi(\mathbf{r}) = f(\mathbf{r})e^{-r/\lambda}$, where the characteristic length $\lambda$ is the localization length and $f(\mathbf{r})$ is some function of space. This localization of the wavefunctions leads to absence of diffusion, with the conductivity $\sigma = e^2 D(\mu)\rho(\mu) \to 0$, where $e$ is the charge of the fermions, $D(\mu)$ is the diffusion coefficient and $\rho(\mu)$ is the density of states at the chemical potential $\mu$. Near the mobility edge we find that the diffusion constant tends to zero and the localization length diverges, as illustrated in Fig. 1.3b. The critical behaviour on either side of the transition is related by hyper-scaling relations [9], and in 3D the critical behaviour is $\lambda \sim |E_c - E|^{-\nu}$ and $D \sim |E - E_c|^{\nu}$, with $0 < \nu < 1$.

Localization is a surprising phenomenon. When we first learn about quantum mechanics we are taught that particles are able to tunnel through classically impenetrable barriers [50]. Take for example the double potential well, where a particle placed in one side can eventually make it to the other side. This can be understood by considering the eigenstates of the two separate wells and those of the combined system. Given the Hamiltonian $\hat{H}$ that describes the system, the eigenstates $|\psi_E\rangle$ are those which are unchanged (up to a real prefactor) under the action of the Hamiltonian, i.e. $\hat{H}|\psi_E\rangle = E|\psi_E\rangle$. By creating symmetric/antisymmetric combinations of the eigenfunctions of separate wells the energy can be increased/lowered, respectively, see Fig. 1.4a. This hybridization of eigenfunctions is typical in coupled quantum systems and seems to conspire against the localization of quantum particles.

In the weak disorder limit there is a simple picture for understanding localization. The idea is to consider multiple different paths through the system between points $A$ and $B$. As the particles travel along these paths they pick up random phases from scattering off the random potential at each site [5, 6, 41, 49]. Due to these random phases, we will get interference between the different paths. More explicitly, let $A_i$ be the probability amplitudes associated with a particular path, then the probability of a particle travelling between sites $A$ and $B$ is given by

$$\Big|\sum_i A_i\Big|^2 = \sum_i |A_i|^2 + \sum_{i\neq j} 2\mathrm{Re}[A_i A_j^*]. \tag{1.22}$$

Due to the random phases from scattering off the potential, we have cancellation in the second term which reduces the probability. If, however, we consider closed paths that start and end at site $A$, as shown in Fig. 1.4b, then for each path there exists a time-reversed counterpart which has the same probability amplitude. These two paths then constructively interfere and we get $|A_1 + A_2|^2 = 4|A_1|^2$. That is, the probability of returning to the same site is enhanced which leads to the localization of particles. Clearly these arguments are too simplistic and a complete understanding requires some more sophistication, but nevertheless, localization is fundamentally a quantum interference effect. Despite their simplicity, these arguments are also powerful in understanding physical phenomena such as magnetoresistence [39, 41–43, 49]. If we include a magnetic field in the above arguments then the original and time reversed closed paths no longer have the same phase and we get constructive or destructive interference depending on the flux through the paths. This leads to either a decrease or increase in the resistivity depending on the magnetic field. If we construct a cylindrical system with the magnetic flux threading the cylinder then the flux through closed electron loops is fixed. This leads to oscillations in the resistance as a function of magnetic field strength [43], see Fig. 1.5a.

One of the most important contributions to understanding Anderson localization came from Abrahams et al. [51]. They used renormalization group methods to deduce the conductance in a disordered system. The central idea is to assume that the

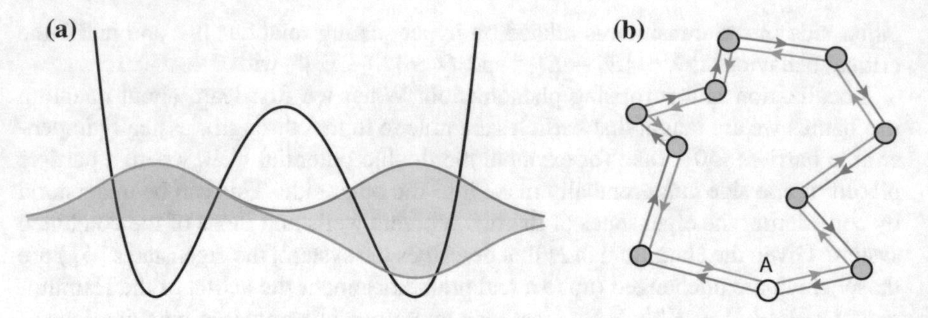


**Fig. 1.4** **a** Schematic of the quantum mechanical double well problem. The two lowest energy states are formed from the symmetric and antisymmetric combinations of the lowest energy single-well states. **b** Two closed paths that are related by time reversal. Grey circles indicate scattering centres, which contribute a random phase

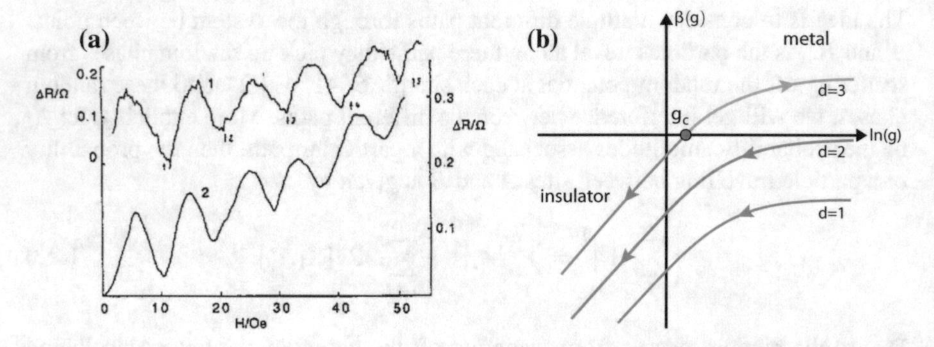


**Fig. 1.5** **a** Result of resistance measurement on a cylindrical geometry, which shows oscillations as a function of the magnetic flux threading the cylinder. The figure is reproduced from the journal Reports on Progress in Physics [49]. **b** Scaling theory for localization, where $d$ is the spatial dimensionality. $\beta(g) > 1$ corresponds to a renormalization group flow towards a conductor whereas $\beta(g) < 1$ flows towards an insulator

dimensionless conductance $g(L)$ scales as a function of the linear length of the system $L$. The renormalization flow equation is

$$\frac{\mathrm{d}\ln g(L)}{\mathrm{d}\ln L} = \beta(g). \tag{1.23}$$

The functional form of $\beta(g)$ then tells us whether for a given disorder strength, which corresponds to a finite conductance in a finite system, the system is a metal or an insulator in the limit $L \to \infty$. The scaling function $\beta(g)$ is shown in Fig. 1.5b. Importantly this shows that the system is always insulating in 1D and that in 3D there exists a critical disorder strength, below which there exists delocalized states and the system is conducting.

We have tried to give a brief overview of the phenomenology and importance of Anderson localization. There is now a very rich history of the subject and a lot is known. We point the reader to a few references for further information [5, 41, 49].

### 1.3.2 *Many-Body Localization*

While Anderson localization is itself an important concept, a major breakthrough was establishing the existence of localization in an interacting many-body quantum system. The resulting many-body localized (MBL) phase represents a robust setting for non-ergodic behaviour in a quantum system [17, 18, 46–48, 52] and has raised fundamental questions about thermalization.

Let us again consider a concrete setting by generalizing the Anderson model of localization (1.21). For example, we consider the Hamiltonian

$$\hat{H} = -J\sum_{\langle jk\rangle}\hat{c}_j^\dagger\hat{c}_k + \sum_j V_j\hat{c}_j^\dagger\hat{c}_j + U\sum_{\langle jk\rangle}\hat{n}_j\hat{n}_k, \tag{1.24}$$

where we have added nearest neighbour density interactions to Eq. (1.21), with $\hat{n}_j = \hat{c}_j^\dagger\hat{c}_j$. Another important example is an XXZ model with disorder, defined by the Hamiltonian

$$\hat{H} = -J\sum_{\langle jk\rangle}\left(\hat{\sigma}_j^x\hat{\sigma}_k^x + \hat{\sigma}_j^y\hat{\sigma}_k^y\right) + \Delta\sum_{\langle jk\rangle}\hat{\sigma}_j^z\hat{\sigma}_k^z + \sum_j V_j\hat{\sigma}_j^z, \tag{1.25}$$

which in 1D is equivalent to Eq. (1.24) by the Jordan-Wigner transformation [20], up to some constants and redefinitions of the parameters. The Jordan-Wigner transformation is given by $\hat{\sigma}_j^z = 2\hat{n}_j - 1$ and $\hat{\sigma}_j^+ = \hat{c}_j^\dagger(-1)^{\sum_{l<j}\hat{n}_l} = (\hat{\sigma}_j^-)^\dagger$, where the Pauli matrices are related to the ladder operators via $\hat{\sigma}_j^x = \hat{\sigma}_j^+ + \hat{\sigma}_j^-$ and $\hat{\sigma}_j^y = i(\hat{\sigma}_j^- - \hat{\sigma}_j^+)$.

The MBL phase has some common properties with Anderson localization, particularly the persistent memory of initial states. This has been studied both theoretically and experimentally. For instance, if we consider Eq. (1.24) then we can start the system in a charge density wave $|\cdots 101010\cdots\rangle$, or in a domain wall $|\cdots 111000\cdots\rangle$. Signatures of this initial state—measured by the density imbalance between even and odd sites, for the charge density wave state; or the difference in number of particles between the two halves of the system for the domain wall—do not reach their thermal value at long times and therefore retain information about the initial state, see Fig. 1.6a.

There are, however, a few characteristic ways in which MBL differs from its non-interacting counterpart. Firstly, unlike Anderson localization, there is a critical disorder strength required to localize particles in interacting systems (in 1D). Secondly, since localized systems retain memory of their initial states it has been proposed that

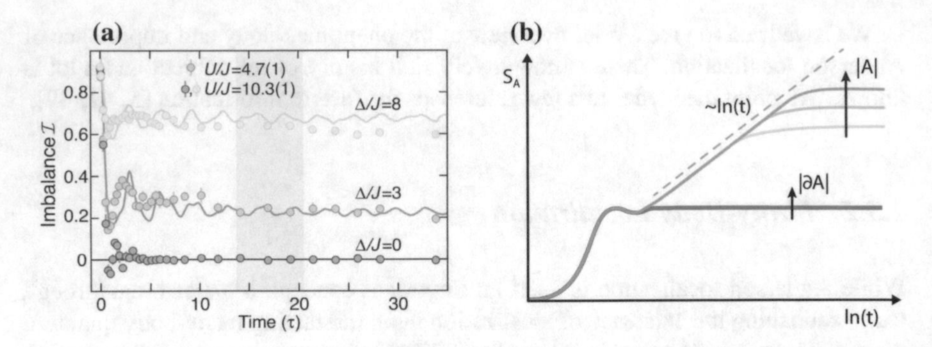


**Fig. 1.6** **a** Experimental results showing the decay of a charge density wave after a quench in an MBL system. The figure is reproduced from the journal Science [53]. The imbalance measures the average difference in density between even and odd sites, which will be zero for a translationally invariant state, but takes a non-zero long-time value in the MBL phase. **b** Behaviour of the entanglement entropy after a global quench in localized systems. The blue curve corresponds to Anderson localization, which after initial linear growth saturates to a value that scales with the area law (constant in 1D). Orange curves correspond to MBL for different partition sizes $|A|$. It displays logarithmic growth until saturation which scales with the size of $|A|$, i.e., a volume law

they could be used in quantum memory devices. However, the interactions in MBL systems result in additional dephasing, which does not occur in Anderson localized systems. A third, and perhaps the most important property of MBL systems, is the robustness to perturbations. It has been demonstrated, and proven in 1D [52], that the MBL phase survives the addition of sufficiently weak, but otherwise generic, additional perturbations. This separates this class of models from Anderson localized and integrable models. In Sect. 1.3.4 we will discuss how MBL systems are closely related to integrable models, and possess an extensive set of "quasi-local" conserved quantities.

### 1.3.3 The Area Law and Logarithmic Entanglement Growth

Localized systems, particularly MBL systems, have some important and distinguishing entanglement properties. Let us first set the scene and define the area and volume law scaling of entanglement. As always we divide our system into $A$ and $B$. Unlike in the context of thermalization we will often simply divide the system into equal parts. We can then discuss how entanglement of a given state, or at a given time after the quench, scales with the size of the partitions $A$ and $B$, which for an equal partition is controlled by the size of the system. For a thermal state, and for a randomly selected state in the Hilbert space, the entanglement entropy scales with the volume of the smaller partition, that is they have volume law scaling. There is, however, a small but important selection of states that do not follow the volume law and instead follow an area law, which is to say they scale with the size of the boundary between $A$ and $B$.

Notably, this latter scaling law is typically obeyed by ground states of local Hamiltonians and can be proven for gapped local Hamiltonians in 1D, see Ref. [54] for more details and references. Furthermore, by the eigenstate thermalization hypothesis, the excited eigenstates of thermalizing systems also themselves look locally thermal and obey a volume law scaling.[6]

Turning to the case of localized systems, both the Anderson and the many-body localized systems have area law scaling for all or a finite fraction of their eigenstates. On the other hand, following a quantum quench, the dynamical behaviour of the entanglement can differentiate the two. Firstly, while in Anderson localized systems the entanglement will grow until saturation which follows area law scaling, MBL systems display logarithmically slow growth, see Fig. 1.6b. In the thermodynamic limit the growth of entanglement is bounded only by the size of the smaller of the two partitions [47, 48], unlike in Anderson localized systems where it is bounded by the size of the boundary between the partitions. All of these properties can be understood when we introduce the $l$-bit description of localized systems in the following section.

### 1.3.4 Local Integrals of Motion (LIOMs) or l-bits

A key insight for understanding MBL was the theory of local integrals of motion (LIOMs), which are commonly referred to as $l$-bits in the context of MBL [55, 56]. In the following discussion we follow closely Ref. [18]. These LIOMs are quasi-local operators $\hat{\tau}_j^z$ which we define as being exponentially close to a local operator—where we reiterate that a local operator is one with finite support in the thermodynamic limit. By exponentially close to a local operator we mean that there exists a local operator $\hat{I}$ with support on a region of size $R$ such that

$$\left\| \hat{\tau}_j^z - \hat{I} \right\| < e^{-R/\xi}, \tag{1.26}$$

where $\| \cdot \|$ denotes some operator norm and $\xi$ is a characteristic length scale for the LIOM.

Let us again be a bit more concrete and consider the spin Hamiltonian (1.25). The LIOMs then take the general form

$$\hat{\tau}_j^z = \hat{\sigma}_j^z + \sum_{\alpha,\beta} \sum_{k,l} f_{kl}^{\alpha\beta} \hat{\sigma}_k^\alpha \hat{\sigma}_l^\beta + \cdots, \tag{1.27}$$

where the coefficients $f_{kl}^{\alpha\beta} \propto \exp\{-|k-l|/\xi\}$ decay exponentially with the distance of the furthest spin from $j$, and similarly for all higher terms in the expansion. The characteristic length scale $\xi$ defines an effective radius of the LIOM. While such an

---

[6] Strictly speaking ETH only implies this volume scaling if $A$ is taken much smaller than $B$ since the density matrices are only locally equivalent.

expansion is always formally possible, the spatial decay of the coefficients is one of the defining features of MBL systems. It is also clear from this expansion that we can truncate this sum such that we include only those terms with support on a region $R$ to get a local operator, and that this operator satisfies Eq. (1.26).

Since these $l$-bits commute with the Hamiltonian, $[\hat{\tau}_j^z, \hat{H}] = 0$, the Hamiltonian can only consist of products of $\hat{\tau}^z$ and takes the general form

$$\hat{H} = \sum_j h_i \hat{\tau}_j^z + \sum_{j,k} J_{jk} \hat{\tau}_j^z \hat{\tau}_k^z + \sum_{j,k,l} J_{jkl} \hat{\tau}_j^z \hat{\tau}_k^z \hat{\tau}_l^z + \cdots, \tag{1.28}$$

where again the coefficients $J_{jk} \propto J_0 \exp\{-|j-k|/\tilde{\xi}\}$, and similarly for higher orders, where $\tilde{\xi}$ is the characteristic length scale for dephasing. Note that in the case of Anderson localization this series expansion would only include the first term and no higher-order products of LIOMs.

An extremely important property of the $l$-bit picture is that it is robust against generic, sufficiently weak, local perturbations to the Hamiltonian (1.25). This is in stark contrast to many integrable systems where the introduction of interactions completely destroys the conservation laws. For MBL, the LIOMs are effectively able to reshuffle in such a way that they remain conserved and quasi-local and the form of Eq. (1.28) is preserved. This robustness has been proven for certain 1D systems in Ref. [52] under a set of reasonable assumptions.

As mentioned above, a characteristic difference between Anderson localization and MBL is the logarithmic growth of entanglement in the latter following a quench, shown in Fig. 1.6. This behaviour can be understood using the $l$-bit form of the Hamiltonian (1.28). For short times the dynamics is controlled by the first term and entanglement is able to build up over regions of the size $\xi$, the effective radius of the LIOMs. For the Anderson localized systems we therefore observe an area law plateau at long times if our partition $A$ is larger than this length scale $\xi$. Since the eigenstates of both Anderson and MBL systems can be labelled by the LIOMs, this length scale also explains the area law entanglement of the eigenstates. The higher order terms present in Eq. (1.28) for MBL, provide a dephasing mechanism between spatially separated $l$-bits and thus lead to the spreading of the entanglement. However, since $\hat{\tau}^z$ are conserved quantities, distant $l$-bits can only become entangled due to direct interactions. This is in contrast to the more generic case where entanglement between spins labelled $A$ and $C$ can be induced by direct interactions between $A$ and an intermediate spin $B$ and between $B$ and $C$. The time scale over which distant $l$-bits can become entangled is set by the inverse coupling strength, i.e.

$$t_{\text{ent}}(r) \sim \frac{1}{J_0 \exp\{-r/\tilde{\xi}\}}, \tag{1.29}$$

implying that entanglement can build up over a region of size $r(t) \sim \tilde{\xi} \ln(J_0 t)$, which grows logarithmically in time.

Using this $l$-bit picture we can provide an explanation for the observed behaviour of the bipartite entanglement entropy. The number of spins that become appreciably entangled with each other after a time $t$ is $N \sim \ln(t)$, and the entanglement with the rest of the spins is exponentially small. Let us suppose that all spins within a radius of $N$ from the boundary between $A$ and $B$ are maximally entangled. This region contains $2NL_A$ spins, where $L_A = |\partial A|$, the size of the boundary of $A$. Since we assume these spins are maximally entangled we have $2^{NL_A}$ non-zero eigenvalues of the reduced density matrix that are all equal.[7] The von Neumann entanglement entropy is then given by

$$S_A = -2^{NL_A}\left(2^{-NL_A}\ln\left(2^{-NL_A}\right)\right) = NL_A \ln 2. \tag{1.30}$$

Given how $N$ depends on time, we end up with $S_A \sim L_A \ln(t)$. We therefore find that the entanglement grows logarithmically in time and that it scales with the size of the boundary between the partitions, as observed numerically and shown in Fig. 6.1b.

## 1.4 Disorder-Free Localization

Early on in the discussion of localization the question was raised whether disorder is a necessary pre-condition, or whether localization can be induced through interactions alone. The work in this thesis provides an answer to this long-standing question by giving the first example of disorder-free localization. One of the first proposals came in 1984 by Kagan and Maksimov [57] in the context of Helium mixtures. This idea was taken up in the last 5 years where multi-species models were introduced [58–60] inspired by Kagan and Maksimov's original idea. We will now give a brief overview of the progress in this area and motivate the problem we address in this thesis.

Kagan and Maksimov's [57] proposal concerned the diffusion of a small concentration of $He^3$ atoms through a regular lattice of solid host $He^4$. The $He^3$ atoms are mobile and interact with each other, and the $He^4$ atoms provide a regular 3D lattice potential but otherwise do not enter the model. More explicitly, the effective model they considered was described by the Hamiltonian

$$\hat{H} = \Delta_0 \sum_{\mathbf{r},\mathbf{n}} \hat{c}^\dagger_{\mathbf{r}+\mathbf{n}}\hat{c}_{\mathbf{r}} + \frac{1}{2}\sum_{\mathbf{r}_1,\mathbf{r}_2} U(\mathbf{r}_1 - \mathbf{r}_2)\hat{c}^\dagger_{\mathbf{r}_1}\hat{c}^\dagger_{\mathbf{r}_2}\hat{c}_{\mathbf{r}_2}\hat{c}_{\mathbf{r}_1}, \tag{1.31}$$

with power law potential $U(r) = U_0(a_0/r)^3$ where $a_0$ is the 'radius' of the unit cell, and $U_0$ is assumed positive.[8] The $\hat{c}$ operators correspond to the creation/annihilation operators for the fermionic $He^3$ atoms.

[7] Note the factor of $1/2$ in the exponent is due to our partition cutting through the centre of this region and so our reduced density matrix covers half of this number of spins.

[8] The condition of positivity is not necessary. In fact the authors of Ref. [57] comment on the form of the physical interaction which changes sign as a function of angle.

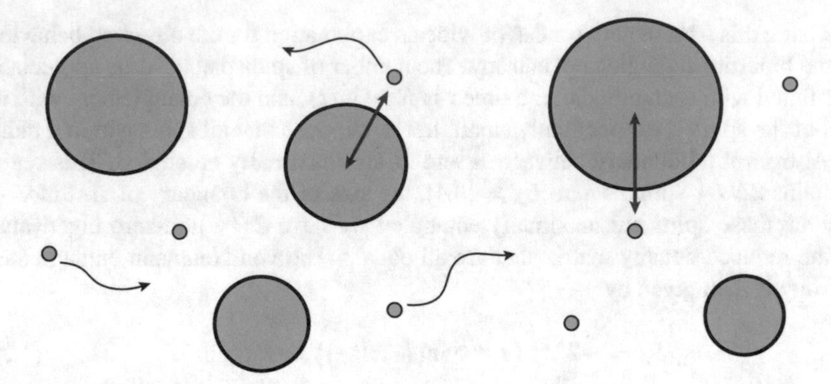

**Fig. 1.7** Cartoon of the scenario considered by Kagan and Maksimov in Ref. [57]. At a critical density the fermions form macroscopic immobile clusters, shown by red circles. The remaining fermions are shown as blue circles and are mobile. The mobile particles feel an effective disorder potential due to the interactions with the clusters

Given this model, they argued that most of the fermions form immobile clusters but there remain a small number that are not involved in these clusters, see Fig. 1.7. Furthermore, since these clusters are seeded on initial inhomogeneities (for instance, due to thermal fluctuations) it may be possible for the mobile particles also to become localized due to this static disordered background. There are three important ingredients in their arguments: (i) that the bandwidth is narrow, i.e., $\Delta_0 z \ll U_0$, where $z$ is the coordination number (6 for a cubic lattice); (ii) the long range, power-law form of the interactions; (iii) that the concentration is low, i.e., $\mu = \frac{1}{N}\sum_i \hat{n}_{\mathbf{r}} \ll 1$, where $N$ is the number of lattice sites. However, this concentration must also be above the critical concentration $\mu_c$ which is necessary for forming macroscopic immobile clusters.

The important step of the argument is the formation of macroscopic immobile clusters which relies on the discrete particle positions. If the above conditions are fulfilled then the likelihood of a resonant transition where the kinetic energy is comparable or greater than the interaction energy is low. This means that particles effectively get stuck in clusters, with the remaining particles that are mobile. Kagan and Maksimov likened this situation to “a swarm of bees in winter hibernation”. The remaining fermions that are not trapped in the clusters then see a disordered background due to interactions with immobile clusters, see Fig. 1.7.

The conclusion of Ref. [57] is that at sufficient concentration of $He_3$ the formation of macroscopic immobile clusters gives rise to the localization of the entire system, including atoms which are not part of the clusters. This analysis is consistent with experimental results on solid Helium where the diffusion coefficient decreases with increased concentration of $He^3$ [61].

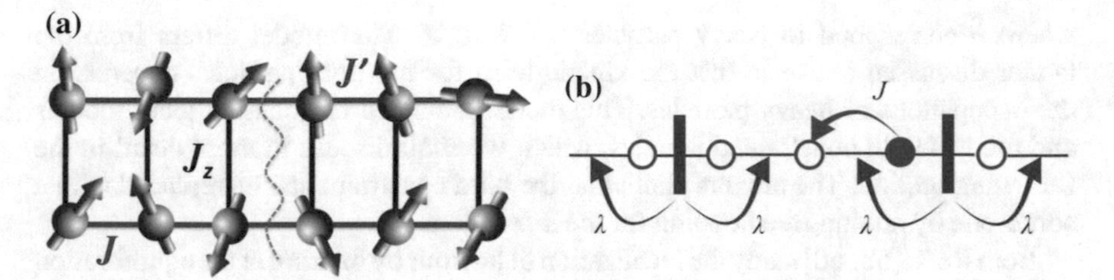


**Fig. 1.8** **a** Schematic of the model in Eq. (1.32). The figure is reproduced from the journal Physical Review Letters [60]. The two chains interact either via XY coupling or Heisenberg exchange and across the rungs the two species are coupled by Ising interactions. **b** The model in Eq. (1.33). The figure is reproduced from the journal AIP conference proceedings [58]. The light fermions live on the odd sites indicated by circles and the heavy fermions live on the even sites and act as hard walls for the light species

### *1.4.1 Heavy-Light Mixtures: Quasi-MBL*

Inspired by work of Kagan and Maksimov, the idea of disorder-free localization was recently taken up again in the form of heavy-light particle models. In these models, the distinction between the 'immobile' and 'mobile' particles is introduced by hand by defining two species of particles with a large mass ratio, and interactions between the two. The question is then whether localization in such systems is possible. These types of models were considered independently by Yao et al. [60] and Schiulaz et al. [58].

The first of the two models we will discuss is the spin ladder model of Yao et al. [60]. The two chains of the ladder are described by XY spin Hamiltonians with different coupling strengths $J$ and $J'$ and an Ising interactions along the rungs of the ladder, see Fig. 1.8a. The model is described by a Hamiltonian

$$\hat{H} = J \sum_{\langle ij \rangle} \hat{S}_i^+ \hat{S}_j^- + J' \sum_{\langle ij \rangle} \hat{s}_i^+ \hat{s}_j^- + J_z \sum_i \hat{S}_i^z \hat{s}_i^z, \tag{1.32}$$

where $\hat{S}$ and $\hat{s}$ are the Pauli operators of the spins on the two chains, where $\hat{S}$ corresponds to the 'heavy' species and thus $J \ll J'$. Reference [60] also considers the case where the XY coupling is replaced by Heisenberg interactions, which shows qualitatively similar behaviour.

Another model was proposed by Schiulaz and Müller [58] and describes two species of mobile fermions on a 1D chain. The light particles hop along the odd sites of a 1D chain and the heavy particles hop along the even sites. Furthermore, the heavy particles act as hard barriers and forbid the hopping of the light particles. The Hamiltonian reads

$$\hat{H} = -\lambda \sum_i \left( \hat{b}_{2i+2}^\dagger \hat{b}_{2i} + \text{H.c} \right) - J \sum_i \left( \hat{a}_{2i+1}^\dagger \hat{a}_{2i-1} + \text{H.c} \right) \left( 1 - \hat{b}_{2i}^\dagger \hat{b}_{2i} \right), \tag{1.33}$$

where $\hat{b}$ correspond to heavy particles and $\lambda \ll J$. This model differs from the ladder discussed above in that the kinetic term for the light particles depends on the occupations of heavy particles. This model is related to quantum glass models and models with constrained kinetics, which we shall discuss in more detail in the following section. The authors claim that the hard constraint can be replaced with a softer one by adding free hopping for the $\hat{a}$ fermions.

Both Refs. [58, 60] study the localization behaviour by looking at the equilibration of an initial state after a quench. Yao et al. [60] consider a spin polarized state with small inhomogeneous spin modulation and measure the average polarization as a function of time after a quench. This polarization is observed to persist indicating localization but at long times the polarization vanishes. Schiulaz et al. [59] reveal a similar story when they start from a random state of local fermion occupation. They measure the average difference in the local density of fermions on neighbouring sites on the heavy sub-lattice. This quantity measures the inhomogeneity of the initial state. This inhomogeneity is found to persist, but again they find that at long times it vanishes.

Importantly, both authors find that the time scale over which information about the initial state persists scales exponentially with the system size. This contrasts the behaviour of MBL where memory of the initial states persists indefinitely. The authors of Ref. [60] thus call this phenomena quasi-MBL. Despite agreement between these two groups, a later paper by Papic et al. [62] highlighted the importance of finite size effects which aid the stability of initial states in small systems. Their results therefore cast doubt on this quasi-MBL behaviour in the thermodynamic limit and suggest that further work is needed.

### *1.4.2 Quantum Glasses and Kinetically Constrained Models*

An alternative approach to heavy-light mixtures is to consider quantum versions of classically glassy models and kinetically constrained quantum models. These models often display slow relaxation in classical dynamics, and quantum analogues have been proposed as disorder-free examples of MBL systems. Here we will restrict ourselves to consider the classical East model [63] and its quantum counterpart [64]. We point the reader to Refs. [63–68] for other examples and further discussions.

We consider first the classical East model [63], which is defined for a string of variables $\{0, 1\}$ with the Hamiltonian

$$\hat{H} = \sum_j \hat{n}_j, \tag{1.34}$$

where $n_j = 0, 1$ reads off the digit at position $j$ in the string such that 1 is associated with an energy cost. The dynamics is then given by the allowed processes $10 \to 11$ and $11 \to 10$, which can generally occur with different probability. These rules encapsulate a dependence on locally changing a digit depending on the digit to its

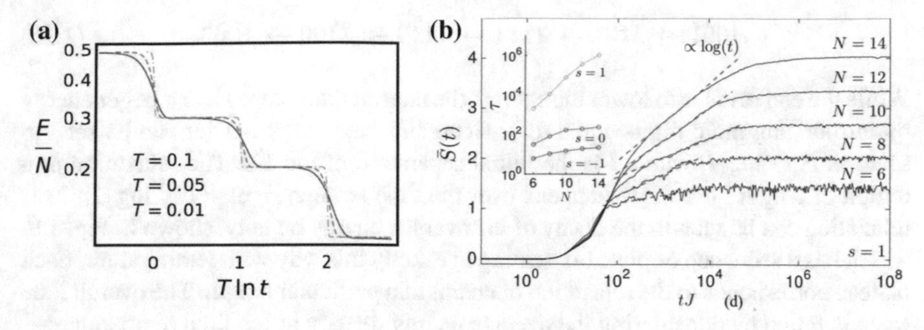


**Fig. 1.9** **a** Relaxation of the energy density due to classical thermal fluctuations following a quench from infinite temperature for the East model (1.34). The results are shown for three values of the final temperature. The figure is reproduced from the journal Physica A: Statistical Mechanics and its Applications [63]. **b** Growth of the entanglement entropy following a global quantum quench in the quantum East model (1.37). Dashed lines indicate the logarithmic fit consistent with MBL. (inset) The relaxation time, $\tau$, extracted from $\bar{S}(t)$, where $s = -\ln \lambda$, see Eq. (1.37). The figure is reproduced from the journal Physical Review B [64]

left. In particular, the dynamics is constrained such that if the digit to the left is zero, then the digit cannot be flipped.

Let us now consider classical dynamics due to thermal fluctuations after a quench from infinite temperature to finite temperature $T$. We prepare the system in a random configuration and then perform evolution using Monte Carlo steps where the dynamics is generated by local digit-flips. The simplest Monte Carlo procedure we can consider is the Metropolis-Hastings algorithm where in each step of the evolution a digit is selected at random and flipped (if allowed by the dynamical rules) to create a new configuration. This new configuration is then accepted with probability $P = \min\left\{1, e^{-\beta(H_{\text{new}} - H_{\text{old}})}\right\}$, where $\beta = T^{-1}$, is the inverse temperature, and $H_{\text{new}}$ and $H_{\text{old}}$ are the energies of the new and old configurations, respectively. If a new configuration is not accepted, we keep the old configuration. The evolution is performed starting from a set of random initial states and then averaged uniformly, which corresponds to a quench from infinite temperature.

The characteristic time scale for the relaxation in this model is given by

$$\tau_{\text{rel}} \sim \exp\left\{\frac{1}{T^2 2 \ln 2}\right\}, \tag{1.35}$$

which scales exponentially with the square of the inverse final temperature. This anomalously slow relaxation[9] can be understood by considering the dynamics required to reduce the energy of a particular string. For instance, consider the string 1001, then a lower energy string can be formed by flipping the far right digit. However, this cannot be done directly because of the constraints, but can be achieved via intermediate states, such as

[9]This slow relaxation is called super-Arrhenius, where Arrhenius is $\sim e^{1/T}$.

$$1001 \rightarrow 1101 \rightarrow 1111 \rightarrow 1110 \rightarrow 1100 \rightarrow 1000. \tag{1.36}$$

While the end result is to lower the energy, the intermediate steps have a larger energy by introducing more digits equal to 1. Generally the energy barrier can be seen to scale as $\Delta \sim \log_2 l$, where $l$ is the initial separation of the 1's. The relaxation of a region of length $l$ therefore happens over the time scale $\tau_l \sim \exp\{T^{-1} \log_2 l\}$. This relaxation can be seen in the decay of the average energy density, shown in Fig. 1.9, which has a sequence of plateaux scaling in exactly this way with temperature. Each plateau corresponds to the relaxation of chains of a particular length. The overall time scale is found by considering the average energy density at the final temperature—i.e., the average separation between ones at the final temperature—which gives an extra factor of $T$ in the denominator of the exponent.

A quantum analogue of the East model was considered in Refs. [63, 64], and is defined by the Hamiltonian

$$\hat{H} = \sum_j \hat{n}_j - \lambda \sum_j \hat{n}_j \hat{\sigma}^x_{j+1}, \tag{1.37}$$

where the binary digits are replaced by spins-1/2 and $\hat{n}_j = \frac{1}{2}(\hat{\sigma}^z_j + 1)$. The first term is the Hamiltonian of the classical model, and the second leads to quantum dynamics where spin-flips are allowed only if the spin is up on the site to the left, in analogy with the East model. The model (1.37) exhibits behaviour consistent with MBL such as the incomplete decay of an inhomogeneous initial state and the logarithmic growth of entanglement entropy, see Fig. 1.9b. Over the time scales accessible to numerical simulations, complete relaxation has not been observed but Ref. [64] could not rule out the possibility of quasi-MBL behaviour.

## 1.5 Lattice Gauge Theories

In this thesis we present a model that realises the physics of disorder-free localization. This model is an example of a lattice gauge theory (LGT) that has a connection to a number of other important LGTs. Gauge theories play a central role in theoretical physics, most famously in the unified description of fundamental particles in the standard model [69]. Gauge theories defined on a lattice also appear increasingly in the description of condensed matter systems, where LGT models often arise as effective descriptions of strongly-correlated systems [20]. In the following sections we discuss a derivation of lattice gauge theories starting from continuum models [19, 70, 71]. The transition from the continuum to the lattice is not only instructive but also introduces the important lattice Schwinger model [70–73] which we shall discuss more later. We review Wegner's Ising gauge theory both for its historical significance and because its derivation is much closer in spirit to many aspects of our model. And finally we consider the example of Kitaev's toric code [68, 74], which is an important model both as a lattice gauge theory and as a prototypical quantum memory. This model is also related to our model of disorder-free localization.

One of the simplest and most familiar gauge theories is classical electrodynamics [75]. Maxwell's equations for the electric and magnetic fields are

$$\nabla \cdot \mathbf{E} = \rho, \qquad \nabla \cdot \mathbf{B} = 0,$$
$$\nabla \times \mathbf{E} = -\frac{\partial \mathbf{B}}{\partial t}, \qquad \nabla \times \mathbf{B} = \mathbf{j} + \frac{\partial \mathbf{E}}{\partial t}, \tag{1.38}$$

where we have set $c = 1$, and $\rho(\mathbf{x}, t)$ and $\mathbf{j}(\mathbf{x}, t)$ are the electric charge and current densities. We are able to introduce a vector potential $\mathbf{A}(\mathbf{x}, t)$ such that $\mathbf{B} = \nabla \times \mathbf{A}$, and if we also introduce a scalar potential $\Phi(\mathbf{x}, t)$, then we can write the electric field as $\mathbf{E} = -\nabla\Phi - \partial_t \mathbf{A}$. The electric and magnetic fields are then completely determined by the scalar and vector potentials $\Phi$ and $\mathbf{A}$.

This representation has a degree of redundancy, which is referred to as gauge freedom or gauge symmetry. This can be parametrized by twice differentiable functions $\alpha(\mathbf{x}, t)$, where the electromagnetic fields are invariant under the transformations

$$\Phi \to \Phi + \partial_t \alpha, \qquad \mathbf{A} \to \mathbf{A} - \nabla\alpha. \tag{1.39}$$

In this classical theory, the introduction of scalar and vector potentials is simply a convenient mathematical description which comes at a cost of this unphysical redundancy.

### 1.5.1 From the Continuum to the Lattice

Let us now consider a continuum quantum theory. We will follow closely Ref. [19] and we also point the reader to Refs. [70, 71]. We take as our starting point the $(3 + 1) - d$ Dirac Lagrangian density

$$\mathcal{L} = \int \mathrm{d}x \; \bar{\psi}(i\gamma^\mu \partial_\mu - m)\psi, \tag{1.40}$$

where $\psi$ are 4 component spinors, the Dirac matrices $\gamma^\mu$ which act on this spinor space, and $\bar{\psi} = \psi^\dagger \gamma^0$. We wish to make this Hamiltonian invariant under the $U(1)$ local gauge transformation to arrive at a theory for quantum electrodynamics (QED) [69]. The $U(1)$ gauge transformation is given by $\psi \to \Lambda(x)\psi$, where $\Lambda(x) = e^{i\alpha(x)}$ and $\alpha(x)$ is a twice differentiable real function.

Now clearly the mass term $m\bar{\psi}\psi$ is gauge invariant since the phase factors between the field and its complex conjugate cancel. However, the kinetic term is not invariant since the derivative also acts on the spatial function $\alpha(x)$ introduced by the transformation. To remedy this we must replace the derivative with covariant derivatives. We will do this by introducing parallel transporters [19]. These may be familiar to the reader from general relativity where the gauge transformations are diffeomorphisms of space-time.

For each space-time curve $\mathcal{C}$ from point to $x$ to $y$, we define the parallel transporters $U(\mathcal{C}_{yx}) : V_x \to V_y$ as maps between the local vector spaces for the fields, i.e. spinor space. That is, $\psi(x) \in V_x$ and $U(\mathcal{C}_{yx})\psi(x) \in V_y$. These parallel transporters respect the group properties of the composition of curves. They allow us to compare distant spinors by first mapping them into the same local vector space by using the information about the function $\alpha(x)$. Under the gauge transformation these parallel transporters transform as

$$U(\mathcal{C}_{yx}) \to \Lambda(y)U(\mathcal{C}_{yx})\Lambda^{-1}(x). \tag{1.41}$$

With this we can define the covariant derivative $D_\mu$ as

$$D_\mu\psi(x) = \lim_{\mathrm{d}x_\mu \to 0} \frac{U(\mathcal{C}_{x,x+\mathrm{d}x_\mu})\psi(x+\mathrm{d}x_\mu) - \psi(x)}{\mathrm{d}x_\mu}. \tag{1.42}$$

If we substitute the normal derivative for this covariant one, then due to the way the parallel transporters transform we have that $D_\mu\psi(x)$ transforms as a spinor field and thus the action is invariant under the gauge transformation. We can expand the parallel transporters along the infinitesimal curves $\mathcal{C}_{x,x+\mathrm{d}x_\mu}$ as $U(\mathcal{C}_{x,x+\mathrm{d}x_\mu}) = \mathbb{1} + A_\mu(x)\mathrm{d}x^\mu$, where the $A_\mu(x) = (\phi(x), -\mathbf{A}(x))$ are real gauge fields that describe the electric and magnetic field. Using this expansion, the covariant derivative takes a more familiar form

$$D_\mu\psi(x) = \left(\partial_\mu + iA_\mu(x)\right)\psi(x). \tag{1.43}$$

From the transformation of the parallel transporters we can see that the gauge field $A_\mu(x)$ transforms as

$$A_\mu \to \Lambda A_\mu \Lambda^{-1} - i\Lambda(\partial_\mu\Lambda^{-1}) = A_\mu - \partial_\mu\alpha, \tag{1.44}$$

just like in classical electrodynamics. The gauge invariant Lagrangian is then

$$\mathcal{L} = \int \mathrm{d}x\ \bar{\psi}(i\gamma^\mu(\partial_\mu + iA_\mu) - m)\psi - \frac{1}{4}F^{\mu\nu}F_{\mu\nu}, \tag{1.45}$$

where in the final term we have given dynamics to the gauge field by promoting it to an operator and using the Yang-Mills Lagrangian where $F_{\mu\nu} = \partial_\mu\hat{A}_\nu - \partial_\nu\hat{A}_\mu$ is the *field strength* tensor [19, 69, 70]. These Yang-Mills terms corresponds to the Hamiltonian energy density $\frac{1}{2}\int \mathrm{d}x\ (\mathbf{E}^2 + \mathbf{B}^2)$, where $\mathbf{E}$ and $\mathbf{B}$ are the electric and magnetic field.

When moving to the lattice we must restrict ourselves to a discrete set of points. The derivatives of the continuum theory then must be replaced. To do so we also use the parallel transporters, not along infinitesimal displacements, but along the bonds connecting sites of the lattice [19]. In terms of the continuum gauge field the parallel transporters along bonds are given by

$$U(y,x) = \exp\left\{\int_{\mathcal{C}_{yx}} \mathrm{d}x^\mu A_\mu(x)\right\}. \tag{1.46}$$

We then define the covariant derivative in analogy with the continuum case as

$$D_\mu \psi(x) = \frac{U(x, x+\vec{\mu})\psi(x+\vec{\mu}) - \psi(x)}{|\vec{\mu}|}, \tag{1.47}$$

where $\vec{\mu}$ is a vector connecting neighbouring sites and $|\vec{\mu}|$ is the lattice spacing. Again the idea is that the parallel transporters provide a gauge invariant way of relating distant operators that can be independently transformed. In writing down the Hamiltonian for the lattice QED there are a number of subtleties, which we will skip and write down the resulting Hamiltonian. See Refs. [19, 70, 71, 73] for more details, but hopefully analogy with the continuum case should make the origin of all these terms clear.

Out of several possible choices we will use staggered fermions [73], which are spinless fermions with the spin information incorporated into the spatial position $x$. The lattice QED is then described by the Hamiltonian

$$\hat{H} = -t\sum_{\langle xy\rangle} s_{xy}(\hat{\psi}_x^\dagger \hat{U}_{xy}\hat{\psi}_y + \text{H.c}) + m\sum_x s_x \hat{\psi}_x^\dagger \hat{\psi}_x + \frac{e^2}{2}\sum_{\langle xy\rangle} \hat{E}_{xy}^2 - \frac{1}{4e^2}\sum_p (\hat{U}_p + \hat{U}_p^\dagger). \tag{1.48}$$

Let us work through the terms in this Hamiltonian and connect them to the above discussion of continuum QED. First, the sign factors $s_{xy}$ and $s_x$, which come from the representation in terms of staggered fermions, replace the $\gamma$-matrices in the Dirac equation. In a general d-dimensional quantum model these are given by $s_x = (-1)^{x_1+\cdots+x_d}$ and $s_{xy} = (-1)^{x_1+\cdots+x_{k-1}}$, where $\langle xy\rangle$ is a bond in the kth direction. Second, we have the link operators $\hat{U}_{xy}$, which are the parallel transporters (1.46). Since the continuum vector field $A_\mu$ no longer has any meaning on a lattice, $\hat{U}_{xy}$ are now the fundamental degrees of freedom.[10] Next, the bond variables $\hat{E}_{xy}$ are the electric field operators which satisfy

$$[\hat{E}_{xy}, \hat{U}_{x'y'}] = \delta_{xx'}\delta_{yy'}\hat{U}_{xy}, \quad \text{and} \quad [\hat{E}_{xy}, \hat{U}_{x'y'}^\dagger] = -\delta_{xx'}\delta_{yy'}\hat{U}_{xy}^\dagger. \tag{1.49}$$

And finally we define the plaquette terms $\hat{U}_p = \hat{U}_{xy}\hat{U}_{yz}\hat{U}_{zw}\hat{U}_{zx}$ as the (oriented) product of link operators around the fundamental square plaquettes of the lattice. These plaquette terms measure the magnetic flux through the plaquette and this term in the Hamiltonian corresponds to $\int \mathrm{d}x\, \mathbf{B}^2$.

---

[10]In an external magnetic field, for example, these bond variables $U_{xy}$ are numbers and correspond to Peierls phases.

Let us also consider the specific case of $(1+1)-$ d QED—the Schwinger model. On the lattice we have the 1D lattice Schwinger model [72, 73] described by the Hamiltonian

$$\hat{H} = -iw\sum_n (\hat{\psi}_n^\dagger \hat{U}_n \hat{\psi}_{n+1} - \text{H.c}) + J\sum_n \hat{E}_n^2 + m\sum_n (-1)^n \hat{\psi}_n^\dagger \hat{\psi}_n, \tag{1.50}$$

where we use shorthand notation $\hat{U}_n = \hat{U}_{n,n+1}$, and similarly for $\hat{E}_n$. Notice that we drop the magnetic term which is not relevant to 1D. We have also redefined constants and the operators $\hat{U}_n$ to conform to the common notation in the literature.

The lattice Schwinger model has an extensive set of conserved quantities associated with the gauge freedom which are

$$\hat{G}_n = E_n - E_{n-1} - \hat{\psi}_n^\dagger \hat{\psi}_n + \frac{1}{2}[1 - (-1)^n]. \tag{1.51}$$

The interpretation of these conserved quantities is as a generalization of Gauss' law $\nabla \cdot \mathbf{E} - \rho = 0$. However, in this case the right hand side does not have to be zero and corresponds to having a non-zero charge density in the vacuum. The model is often restricted to act on the states that satisfy the Gauss' law, $\hat{G}_n|\psi\rangle = 0$, i.e., only gauge invariant states.

### 1.5.2 Wegner's Ising Lattice Gauge Theory

An earlier construction of lattice gauge theories came from Wegner [76, 77]. The goal here was to construct a Hamiltonian of Ising variables that obeyed a particular local gauge transformation. He considered both the statistical mechanics problem as well as a quantum Hamiltonian, and the connections between them. Here we focus on the quantum Hamiltonian. This is a model of Ising spins with a local discrete $\mathbb{Z}_2$ gauge symmetry, and is thus called the Ising gauge theory (IGT). Here we briefly consider the 2D IGT as it is relevant for the following discussions. Higher-dimensional generalizations follow naturally.

Let us consider a 2D square lattice with Ising spins on the bonds, see Fig. 1.10. We then want to write down a Hamiltonian invariant under the action of a local operator $\hat{G}_j$, which flips all spins on the bonds connected to site $j$, as illustrated in Fig. 1.10b. This invariance can be written as $\hat{G}_j^{-1}\hat{H}\hat{G}_j = \hat{H}$, where we note that $\hat{G}_j^{-1} = \hat{G}_j$. For Ising spins we can consider two local operators $\hat{\sigma}_{jk}^z$ and $\hat{\sigma}_{jk}^x$, which measure and flip spins on the bond between sites $j$ and $k$. The action of the local operators $\hat{G}_j$ is given as follows

$$\hat{G}_j \hat{\sigma}_{kl}^x \hat{G}_j = \hat{\sigma}_{kl}^x, \qquad \hat{G}_j \hat{\sigma}_{kl}^z \hat{G}_j = (-1)^{\delta_{jk}+\delta_{jl}} \hat{\sigma}_{kl}^z. \tag{1.52}$$

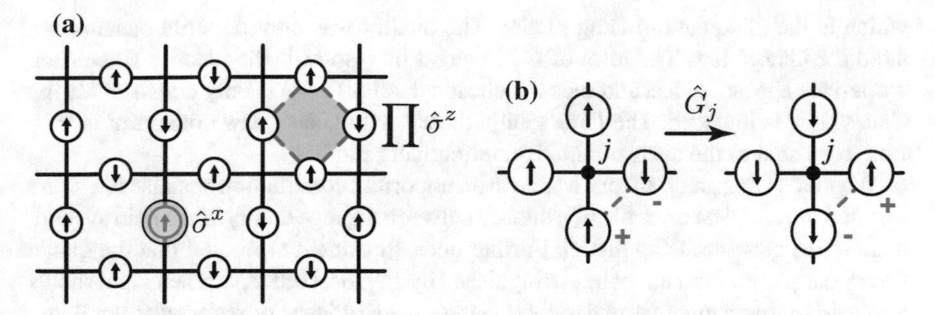


**Fig. 1.10** **a** Schematic picture of Wegner's Ising gauge theory with Ising spins living on the bonds of a square lattice. The two terms in the Hamiltonian are shown as blue circles for $\hat{\sigma}^x$ and the plaquette terms are shown in red. **b** Action of the generators of the gauge symmetry $\hat{G}_j$, which flip all spins on bonds connected to that site $j$. The product of any two $\hat{\sigma}^z$ connected to site $j$ is left invariant which is indicated by the signs shown in red (color figure online)

Therefore, any operator involving any combination of $\hat{\sigma}^x_{jk}$ is automatically gauge invariant. On the other hand, the gauge invariant combinations of $\hat{\sigma}^z_{jk}$ must consist of products around closed loops on the square lattice. This is to ensure that if we have $\hat{\sigma}^z_{jk}$ on a bond connected to site $j$, then we must have another $\hat{\sigma}^z_{jl}$ connected to site $j$ to cancel the sign. The simplest such operator is a product of $\hat{\sigma}^z$ around irreducible square plaquettes of the lattice. The Hamiltonian is

$$\hat{H} = -\sum_{\langle jk \rangle} \hat{\sigma}^x_{jk} - \lambda \sum_{\text{plaquettes p}} \prod_{\langle jk \rangle \in \Box_p} \hat{\sigma}^z_{jk}. \tag{1.53}$$

Note that (1.53) is similar to the transverse field Ising model Hamiltonian but here the Ising coupling has been replaced by the plaquette terms, shown schematically in Fig. 1.10a.

Finally, as well as ensuring that the Hamiltonian is gauge invariant, we must impose the constraint that the physical Hilbert space consists only of states that are invariant under the symmetry action, i.e. $\hat{G}_j|\psi\rangle = |\psi\rangle$, for all $j$. This is analogous to Gauss' law. Given this constraint, we can then perform a duality transformation by defining new spin operators that live at the centres of plaquettes

$$\hat{\tau}^x_p = \prod_{\langle jk \rangle \in \Box_p} \hat{\sigma}^z_{jk}, \qquad \hat{\tau}^z_p \hat{\tau}^z_q = \hat{\sigma}^x_{jk}, \tag{1.54}$$

where $p$ and $q$ are the plaquettes that share a bond $\langle jk \rangle$. After this duality transformation the Hamiltonian takes the form

$$\hat{H} = -\sum_{\langle pq \rangle} \hat{\tau}^z_p \hat{\tau}^z_q - \lambda \sum_p \hat{\tau}^x_p, \tag{1.55}$$

which is the 2D quantum Ising model. The duality was only possible because we fixed the Gauss' law. The form of $\hat{G}_j$ in terms of $\tau$ spins is the identity since each $\tau$ operator appears twice, and thus the chosen duality (1.54) is only consistent if the Gauss' law is imposed. The Ising gauge theory with Gauss' law constraint is then said to be dual to the unconstrained quantum Ising model.

Wegner's Ising gauge theory was such an important contribution because it was the first and the simplest example of a duality between a gauge theory and a spin system, namely the quantum Ising model. Furthermore, this duality revealed that the gauge theory has phases that cannot be distinguished by any local order parameter and shows a transition that cannot be understood using standard ideas of symmetry breaking. In the quantum Ising model at zero temperature we have two phases (ferromagnetic and paramagnetic) depending on the magnitude of $\lambda$, which can be distinguished by a local order parameter—the total magnetization. The order parameters for the corresponding phases in the IGT are necessarily non-local and there is no symmetry breaking. The lack of symmetry breaking is a consequence of Elitzur's theorem which states that a local gauge symmetry cannot be spontaneously broken. The phases of the quantum Ising model, which can be distinguished by the local magnetization, are dual to the phases of the corresponding Ising gauge theory and local order parameters correspond to non-local objects in the gauge theory. The idea of non-local order parameters and lack of spontaneous symmetry breaking is an important part of our understanding of topological phases of matter, an example of which we will discuss in the following section.

### *1.5.3 Kitaev's Toric Code*

The IGT is not the only Hamiltonian we can write down that is invariant under the symmetry action (1.52). Let us consider another important example—Kitaev's toric code [68, 74]. This is formed by replacing the transverse field $\hat{\sigma}^x$, by a product of $\hat{\sigma}^x$ on the bonds connected to a single site of the lattice. Furthermore, the model is defined on a 2D square lattice with periodic boundary conditions. The toric code Hamiltonian is

$$\hat{H}_{TC} = -K \sum_j \hat{A}_j - K' \sum_p \hat{B}_p, \tag{1.56}$$

where we have defined

$$\hat{A}_j = \prod_{k \in +_j} \hat{\sigma}^x_{jk} \qquad \hat{B}_p = \prod_{\langle jk \rangle \in \square_p} \hat{\sigma}^z_{jk}, \tag{1.57}$$

the star and plaquette operators, respectively, see Fig. 1.11a. Note that the star operators, $\hat{A}_j = \hat{G}_j$, are the generators of the gauge symmetry (1.52). The model can be solved since all of the operators in the Hamiltonian commute with one another. This means that all $A_j, B_p = \pm 1$ are good quantum numbers which one can use to

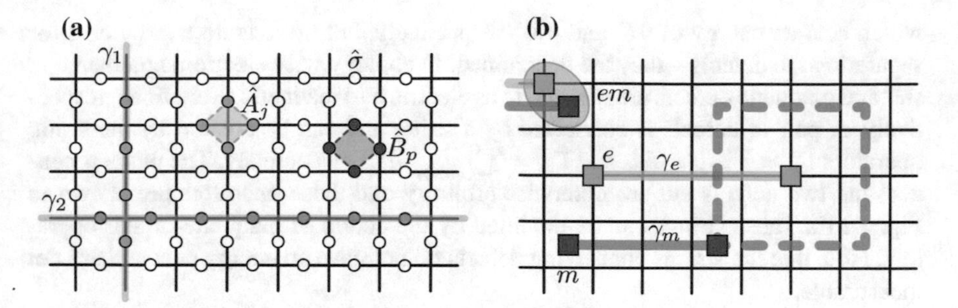


**Fig. 1.11** **a** Schematic picture of Kitaev's toric code model defined on a square lattice with spins $\hat{\sigma}$ on the bonds (white circles). The star, $\hat{A}_j$, and plaquette operators, $\hat{B}_p$, defined in Eq. (1.57) are shown in blue and red, respectively. Loops $\gamma_1$ and $\gamma_2$ are shown in grey. **b** Three types of anyons in the toric code: electric (blue), magnetic (red) and composite (grey ellipse). The solid coloured lines show strings $\gamma_e$ and $\gamma_m$ connecting pairs of defects. The mutual statistics is probed by dragging one defect around another, shown by the dashed line

label the eigenstates. The ground state of the model is given by $\hat{A}_j|GS\rangle = |GS\rangle$ and $\hat{B}_p|GS\rangle = |GS\rangle$, for all stars and plaquettes.

If the system has $N$ sites then there are $2^{2N}$ states but since $\prod_{\text{all}} \hat{A}_j = \prod_{\text{all}} \hat{B}_p = 1$ on the torus, these correspond to $2^{2(N-2)}$ labels. There are therefore 4 degenerate sectors[11], which can't be distinguished by $A_j$ and $B_p$. In fact, these sectors can't be distinguished by any local operators, just like the phases of Wegner's IGT. To distinguish these four sectors we must use non-local Wilson loop operators, which are products of spin operators around non-contractible loops of the torus. For instance we can take two independent loop operators to be

$$\hat{\Gamma}_1 = \prod_{\langle jk\rangle \in \gamma_1} \hat{\sigma}^z_{jk}, \qquad \hat{\Gamma}_2 = \prod_{\langle jk\rangle \in \gamma_2} \hat{\sigma}^z_{jk}, \tag{1.58}$$

where $\gamma_1$ and $\gamma_2$ are arbitrary closed loops that wind once around the two directions of the torus, see Fig. 1.11a. Both Wilson loop operators $\hat{\Gamma}_\alpha$ commute with all $\hat{B}_p$ because they are both products of $\hat{\sigma}^z$ and with all $\hat{A}_j$ because the loop operator will contain a product with only an even number of bonds in a star. These operators also have eigenvalues $\pm 1$ and completely resolve the degeneracy. The four degenerate ground states of the model are topologically protected against perturbations, i.e., they are robust against sufficiently weak but generic local perturbations. For this reason it may be possible to use realisations of the toric code to store quantum information [74].

Another important aspect of this model is the properties of the excitations above the ground state manifold. These excitations are defective stars or plaquettes for which $\hat{A}_j|\psi\rangle = -|\psi\rangle$ or $\hat{B}_p|\psi\rangle = -|\psi\rangle$. These defects must be created in pairs,

[11] More generally if the toric code is defined on a surface with genus $g$ then the degeneracy is $2^{2g}$ [68]. The fact, that the ground state degeneracy is dependent on the topology is a characteristic property of topologically ordered systems.

which cost an energy of $4K$ and $4K'$, respectively, but there is no energy cost for separating the defects—they are deconfined. In analogy with electromagnetism, the star and plaquette defects are referred to as electric and magnetic excitations, respectively. A pair of defects is connected by a string and can be created by the string operators $\hat{\Gamma}_e = \sum_{\langle jk\rangle \in \gamma_e} \hat{\sigma}^x_{jk}$ and $\hat{\Gamma}_m = \sum_{\langle jk\rangle \in \gamma_m} \hat{\sigma}^z_{jk}$, respectively. The paths $\gamma$ connect the two defects but are otherwise arbitrary and these operators are shown in Fig. 1.11b. These strings can be modified by the action of plaquette or star operators. Note that the strings connecting defects do not have an energy cost and are not measurable.

The defects in the toric code have non-trivial mutual statistics due to the presence of these connecting strings. To see this let us consider a pair of defects of each type, as shown in Fig. 1.11b, connected by some arbitrary string between them. Let us then take one of the magnetic defects in a complete loop around one of the electric defects, as indicated by the dashed line in Fig. 1.11b. The final position of all defects is the same as to start with, but since two strings now cross we pick up a negative sign to the wavefunction. This can most easily be seen by progressively contracting the loop by acting with star operators within the loop. This gives a factor of $+1$ for each star except at the defect which gives the overall factor of $-1$. This factor of $-1$ means that the electric and magnetic defects have mutual semionic statistics, but have bosonic exchange statistics. We can further form a composite electric and magnetic defect, as illustrated in Fig. 1.11b, which has mutual semionic statistics with both the electric and magnetic defects but fermionic exchange statistics. The phenomenon of emergent point-like particles with statistics that do not reflect the underlying constituent degrees of freedom is called fractionalization and is common in topologically ordered states of matter.

## 1.6 Experimental Progress

Our work comes at a time when there is exceptional progress in the control of isolated quantum systems. These advances are important for the study of many-body localization and lattice gauge theories, but also more generally are providing access to quantum dynamics beyond what can be simulated numerically.

In recent years there have been several advancements in experiments with cold atomic gases that are particularly relevant for us. These experiments are now quite mature and have been developed over the last 30 years. One of the most notable landmarks in this field was the 1995 observation of Bose-Einstein condensation in a cold atomic gas [78]. These experiments required remarkable new levels of sophistication in the trapping and subsequent laser and evaporative cooling of approximately two thousand rubidium atoms to 170 nano-Kelvin. A common tool in modern experiments is the use of optical lattices, which trap atoms at a predefined set of points. In this setting it is possible to engineer coupling between different atoms, for instance by Feshbach resonance, and also directly image and measure quantum systems, for example, using quantum gas microscopy [79, 80]. These technological capabilities mean that such experiments are able to directly realise lattice Hamiltonians with great precision and tunability and over large time scales. In many cases these systems are strongly correlated and far beyond what is possible to simulate classically. We will mention below in a little more detail a couple of experiments studying many-body localization that are particularly relevant to this thesis.

The experiments we discuss are reminiscent of Feynman's vision of 'quantum simulation' [81], that is, to study the dynamics of quantum system by building another well-controlled quantum system that simulates it. These advances and this vision are by no means restricted to cold atom experiments. On the contrary, as just a few examples we can also mention photonic quantum circuits [82], superconducting chips [83], and quantum simulators with Rydberg atoms [84, 85]. An example that we will consider in a little more detail below is trapped ion experiments, which have recently been used to digitally simulate a quantum lattice gauge theory [86, 87].

While mentioning the advances in the control of quantum systems and quantum simulation it would be remiss of us not to mention the current progress towards universal quantum computation. This field is currently receiving a massive push and support from companies such as IBM, Google, Microsoft, as well as many universities around the world. In principle these devices would allows us to study quantum phenomena beyond the reach of classical computation, but also hold promise for quantum cryptographic and communication protocols. These goals are pushing both technological and theoretical boundaries and rely on unparalleled control of coherent quantum states.

### *1.6.1 Cold Atoms in Optical Lattices*

Let us mention in a little more detail a couple of cold atom experiments that study many-body localization physics. In these experiments a lattice is created using the interference of laser light reflected off mirrors to form a standing wave. The electromagnetic field due the lasers induces a Stark effect on the atoms, meaning that the atoms feel an effective periodic potential trapping them at the nodes of this standing wave. By different combination of lasers at different frequencies, many lattices can be created in one, two or three-dimensions. The atoms trapped in these lattices then have hopping amplitudes due to quantum tunneling and natural on-site interactions. Interactions can also be tuned, for example, by Feshbach resonance. In this way many local Hamiltonians can be realised for both fermionic and bosonic atoms [88].

The first experiment we will discuss is that of Schreiber et al. [53], where they study a 1D generalized Aubry-Andre model with interactions described by the Hamiltonian

$$\hat{H} = -J \sum_{i,\sigma=\uparrow,\downarrow} \left(\hat{c}^{\dagger}_{i+1,\sigma}\hat{c}_{i,\sigma} + \text{H.c}\right) + \Delta \sum_{i,\sigma=\uparrow,\downarrow} \cos(2\pi\beta i + \phi)\hat{n}_{i,\sigma} + U \sum_{i} \hat{n}_{i,\uparrow}\hat{n}_{i,\downarrow}, \tag{1.59}$$

where $\beta$ is irrational and the particles are fermions. The Aubry-Andre model is characterized by a quasi-random on-site potential that is periodic but with incommensurate wavelength to the lattice, which is ensured by taking $\beta$ irrational. Without interactions $U$, the Aubry-Andre model has all states localized above the same critical disorder strength $\Delta/J = 2$ for nearly all irrational values of $\beta$ [89]. This contrasts the truly disordered situation where we have all states localized for arbitrarily weak disorder in 1D. The quasi-periodic potential is implemented in the experiment using two laser frequencies—one with a shorter base wavelength superimposed with a weaker, incommensurate wavelength, which results in a quasi-periodic potential.

The experiment then studies dynamics after a global quantum quench starting from a charge density wave initial state. This initial state is created by trapping and cooling the atoms in an initially large depth lattice with twice the wavelength of the base wavelength. In the actual preparation a given even site can be empty or have single or double occupancy, but with an average occupancy of one in those sites at the bottom of this preparation potential and zero for those at the top. Over an experimentally short time-scale, the system is then ramped to Eq. (1.59) and the charge density wave pattern decays. The decay of the CDW is quantified by the imbalance

$$\mathcal{I} = \frac{N_e - N_o}{N_e + N_o}, \tag{1.60}$$

where $N_{e/o}$ are the total number of particles on even/odd sites respectively, with the denominator included to account for particle loss. The details of how this imbalance is measured is given in the supplemental material for Ref. [53]. These experiments demonstrate a many-body localized regime where the density imbalance decay is incomplete, as shown in Fig. 1.6a. They are also able to map out the phase diagram as a function of the Hamiltonian parameters $\Delta/J$ and $U/J$. More recently, a similar experiment has been carried out on a 2D version of the Hamiltonian (1.59) in Ref. [90].

The second experiment we mention studies many-body localization of bosons in 2D using a quantum gas microscope [80]. The benefits of the quantum gas microscope in this experiment are two-fold: it is used to project an almost site resolved, truly random disorder potential felt by the atoms; and it is used for direct measurement of the site resolved occupation. They simulate a 2D Bose-Hubbard model with on-site disorder described by the Hamiltonian

$$\hat{H} = -J \sum_{\langle jk \rangle} \hat{a}_j^\dagger \hat{a}_k + \frac{U}{2} \sum_i \hat{n}_i (\hat{n}_i - 1) + \sum_i \delta_i \hat{n}_i, \tag{1.61}$$

where both the tunnelling amplitude $J$ and the interaction strength $U$ are fixed. The size of the system is approximately $30 \times 30$ lattice sites, but the effective system size is a little smaller due to the trapping potential and a typical realization will contain $\sim$125 particles. Nevertheless, these systems are far beyond what is possible to simulate using exact diagonalization methods.

The setup that they consider is a global quantum quench protocol with half of the system filled and half empty creating a domain wall. This is then left to evolve under the full Hamiltonian and the remaining imbalance between the two sides provides a signature of the memory about the initial state. This imbalance is defined as

$$\mathcal{I} = \frac{N_L - N_R}{N_L + N_R}, \tag{1.62}$$

where $N_{L/R}$ is the number of particles in the left/right half of the system, respectively. The experiment is able to observe the incomplete melting of the domain wall for sufficiently large values of disorder strength and locate the localization transition as a function of disorder strength and filling.

### *1.6.2 Trapped Ion Quantum Simulators*

The last experiment we will consider consists of an array of trapped ions, which are used for the digital quantum simulation of a Hamiltonian [87]. Unlike in cold atom experiments, the time evolution is not achieved through the real time evolution of the system but through the application of quantum gates that realise a Trotter decomposition of the dynamics. Despite this apparent downside of not being able to continuously simulate dynamics, the advantage lies in the ability to address physically separate ions and implement long range Hamiltonians. In this case the authors simulate the quantum lattice gauge theory of the 1D lattice Schwinger model defined in Eq. (1.50), see also [72, 86].

The important theoretical step is to first rewrite the Hamiltonian using the conserved quantities associated with the gauge symmetry and implement a Jordan-Wigner transformation of the fermions. This results in a spin-1/2 Hamiltonian with at most two-spin interactions. The details of this transformation can be found in the appendix of Ref. [87]. The resulting Hamiltonian has the form

$$\hat{H} = \frac{m}{2}\sum_{n=1}^{N}(-1)^n\hat{\sigma}_n^z + w\sum_{n=1}^{N-1}\left[\hat{\sigma}_n^+\hat{\sigma}_{n+1}^- + \text{H.c}\right] + J\sum_{n=1}^{N-1}\left[\epsilon_0 + \frac{1}{2}\sum_{m=1}^{n}\left[\hat{\sigma}_m^z + (-1)^m\right]\right]^2. \tag{1.63}$$

The simulation is then implemented by Trotter decomposition into local gates and Mølmer-Sørensen (MS) gates which are of the form

$$\hat{H}_{MS} = J_0\sum_{n,m}\hat{\sigma}_n^x\hat{\sigma}_m^x. \tag{1.64}$$

The ions can then be selectively addressed such that only chosen spins are involved in this sum. Combining these with local unitary gates, the full Trotter time evolution can be digitally simulated.

## References

1. Feynman RP (1972) Statistical mechanics: a set of lectures. Westview Press
2. Lifshitz EM, Pitaevskii LP (1980a) Statistical physics part I, 3rd edn. Pergamon Press
3. Lifshitz EM, Pitaevskii LP (1980b) Statistical physics part 2: theory of the condensed state, 2nd edn. Pergamon Press
4. Blundell SJ, Blundell KM (2010) Concepts in thermal physics, 2nd edn. Oxford University Press, Oxford
5. Efetov K (1997) Supersymmetry in disorder and chaos. Cambridge University Press, Cambridge. https://doi.org/10.1017/CBO9780511573057
6. Coleman P (2015) Introduction to many-body physics. Cambridge University Press, Cambridge. https://doi.org/10.1017/CBO9781139020916
7. Altland A, Simons B (2010) Condensed matter field theory, 2nd edn. Cambridge University Press, Cambridge. https://doi.org/10.1017/CBO9780511789984
8. Landau LD (1937) On the theory of phase transitions. Zh Eks Teor Fiz 7:19–32
9. Cardy J (1996) Scaling and renormalization in statistical physics. Cambridge University Press, Cambridge. https://doi.org/10.1017/CBO9781316036440
10. Lifshitz EM, Pitaevskii LP (1981) Physical kinetics. Pergamon Press
11. Keldysh LV (1965) Diagram technique for non-equilibrium processes. JETP 20:1018. http://www.jetp.ac.ru/cgi-bin/e/index/e/20/4/p1018?a=list
12. Kamenev A (2011) Field theory of non-equilibrium systems. Cambridge University Press, Cambridge. https://doi.org/10.1017/CBO9781139003667
13. Berry M (1989) Quantum chaology, not quantum chaos. Phys Scr 40: 335–336. https://doi.org/10.1088/0031-8949/40/3/013
14. Deutsch JM (1991) Quantum statistical mechanics in a closed system. Phys Rev A 43: 2046–2049. https://doi.org/10.1103/PhysRevA.43.2046
15. Srednicki M (1994) Chaos and quantum thermalization. Phys Rev E 50: 888–901. https://doi.org/10.1103/PhysRevE.50.888
16. Deutsch JM (2018) Eigenstate Thermalization Hypothesis. http://arxiv.org/abs/1805.01616 arXiv:1805.01616
17. Nandkishore R, Huse DA (2015) Many-body localization and thermalization in quantum statistical mechanics. Ann Rev Condens Matter Phys 6:15–38. https://doi.org/10.1146/annurev-conmatphys-031214-014726
18. Abanin DA, Papić Z (2017) Recent progress in many-body localization. Ann Phys 529:1700169. https://doi.org/10.1002/andp.201700169
19. Montvay I, Münster G (1994) Quantum fields on a lattice. Cambridge University Press, Cambridge. https://doi.org/10.1017/CBO9780511470783
20. Fradkin E (2013) Field theories of condensed matter physics. Cambridge University Press, Cambridge. https://doi.org/10.1017/CBO9781139015509
21. Wen X-G (2007) Quantum field theory of many-body systems. Oxford University Press, Oxford. https://doi.org/10.1093/acprof:oso/9780199227259.001.0001
22. Essler FHL, Fagotti M (2016) Quench dynamics and relaxation in isolated integrable quantum spin chains. J Stat Mech Theory Exp 2016:064002. https://doi.org/10.1088/1742-5468/2016/06/064002
23. Vasseur R, Moore JE (2016) Nonequilibrium quantum dynamics and transport: from integrability to many-body localization. J Stat Mech Theory Exp 2016:064010. https://doi.org/10.1088/1742-5468/2016/06/064010
24. Gogolin AO, Nersesyan AA, Tsvelik AM (1998) Bosonization and strongly correlated systems. Cambridge University Press, Cambridge
25. Kitaev AY (2006) Anyons in an exactly soled model and beyond. Ann Phys (N Y) 321:2–111. https://doi.org/10.1016/j.aop.2005.10.005
26. Baskaran G, Mandal S, Shankar R (2007) Exact results for spin dynamics and fractionalization in the Kitaev model. Phys Rev Lett 98:247201. https://doi.org/10.1103/PhysRevLett.98.247201

27. Knolle J, Kovrizhin DL, Chalker JT Moessner R (2014) Dynamics of a two-dimensional quantum spin liquid: signatures of emergent Majorana fermions and fluxes. Phys Rev Lett 112:207203. http://journals.aps.org/prl/supplemental/10.1103/PhysRevLett.112.207203
28. Banerjee A, Yan J, Knolle J, Bridges CA, Stone MB, Lumsden MD, Mandrus DG, Tennant DA, Moessner R, Nagler SE (2017) Neutron scattering in the proximate quantum spin liquid $\alpha$-RuCl 3. Science 356:1055–1059. https://doi.org/10.1126/science.aah6015
29. Smith A, Knolle J, Kovrizhin DL, Chalker JT, Moessner R (2015) Neutron scattering signatures of the 3D hyperhoneycomb Kitaev quantum spin liquid. Phys Rev B 92:180408. https://doi.org/10.1103/PhysRevB.92.180408
30. Smith A, Knolle J, Korizhin DL, Chalker JT, Moessner R (2016) Majorana spectroscopy of three-dimensional Kitaev spin liquids. Phys Rev B 93:235146. https://doi.org/10.1103/PhysRevB.93.235146
31. Lieb EH, Robinson DW (1972) The finite group velocity of quantum spin systems. Commun Math Phys 28:251–257. https://doi.org/10.1007/BF01645779
32. Bravyi S, Hastings MB, Verstraete F (2006) Lieb-Robinson bounds and the generation of correlations and topological quantum order. Phys Rev Lett 97:1–4. https://doi.org/10.1103/PhysRevLett.97.050401
33. Srednicki M (2012) KITP talk: overview of eigenstate thermalization hypothesis. http://online.kitp.ucsb.edu/online/qdynamics12/srednicki/
34. Khatami E, Pupillo G, Srednicki M, Rigol M (2013) Fluctuation-dissipation theorem in an isolated system of quantum dipolar bosons after a Quench. Phys. Rev Lett 111:050403. https://doi.org/10.1103/PhysRevLett.111.050403
35. D'Alessio L, Kafri Y, Polkonikov A, Rigol M (2016) From quantum chaos and eigenstate thermalization to statistical mechanics and thermodynamics. Adv Phys 65:239–362. https://doi.org/10.1080/00018732.2016.1198134
36. Wilde MM (2017) Quantum information theory, 2nd edn. Cambridge University Press, Cambridge. https://doi.org/10.1017/9781316809976
37. Eisert J (2001) Entanglement in quantum information theory. Ph.D. thesis. https://arxi.org/pdf/quant-ph/06102531.pdf
38. Li H, Haldane FDM (2008) Entanglement spectrum as a generalization of entanglement entropy: identification of topological order in Non-Abelian fractional quantum hall effect states. Phys Rev Lett 101:010504 . https://doi.org/10.1103/PhysRevLett.101.010504
39. Geraedts SD, Nandkishore R, Regnault N (2016) Many-body localization and thermalization: insights from the entanglement spectrum. https://doi.org/10.1103/PhysRevB.93.174202 Phys Rev B 93:174202
40. Anderson PW (1958) Absence of diffusion in certain random lattices. Phys Rev 109:1492–1505. https://doi.org/10.1103/PhysRev.109.1492
41. Abrahams E ed (2010) 50 years of Anderson localization. World Scientific
42. Altshuler BL, Arono AG, Larkin AI, Khmelnitskil DE, Konstantinou BP (1981) Anomalous magnetoresistance in semiconductors. JETP 54:411. http://www.jetp.ac.ru/cgi-bin/dn/e_054_02_0411.pdf
43. Aronov AG, Sharvin Y (1987) Magnetic flux effects in disordered conductors. Rev Mod Phys 59:755–779. https://doi.org/10.1103/RevModPhys.59.755
44. Bergmann G (1984) Weak localization in thin films. Phys Rep 107:1–58. https://doi.org/10.1016/0370-1573(84)90103-0
45. Tong D (2016) Lecture notes on: the quantum hall effect. http://www.damtp.cam.ac.uk/user/tong/qhe.html
46. Basko DM, Aleiner IL, Altshuler BL (2006) Metalinsulator transition in a weakly interacting many-electron system with localized single-particle states. Ann Phys (N Y) 321:1126–1205. https://doi.org/10.1016/j.aop.2005.11.014
47. Žnidarič M, Prosen T, Prelovšek P (2008) Many-body localization in the Heisenberg XXZ magnet in a random field. Phys Rev B 77:064426. https://doi.org/10.1103/PhysRevB.77.064426
48. Bardarson JH, Pollmann F, Moore JE (2012) Unbounded growth of entanglement in models of many-body localization. Phys Rev Lett 109:017202. https://doi.org/10.1103/PhysRevLett.109.017202

49. Kramer B, MacKinnon A (1993) Localization: theory and experiment. Rep Prog Phys 56:1469–1564. https://doi.org/10.1088/0034-4885/56/12/001
50. Landau LD, Lifshitz EM (1977) Quantum mechanics: non-relativistic theory, 3rd edn. Pergamon Press
51. Abrahams E, Anderson PW, Licciardello DC, Ramakrishnan TV (1979) Scaling theory of localization: absence of quantum diffusion in two dimensions. Phys Rev Lett 42:673–676. https://doi.org/10.1103/PhysRevLett.42.673
52. Imbrie JZ (2016) On many-body localization for quantum spin chains. J Stat Phys 163:998–1048. https://doi.org/10.1007/s10955-016-1508-x
53. Schreiber M, Hodgman SS, Bordia P, Luschen HP, Fischer MH, Vosk R, Altman E, Schneider U, Bloch I (2015) Observation of many-body localization of interacting fermions in a quasirandom optical lattice. Science 349:842–845. https://doi.org/10.1126/science.aaa7432
54. Eisert J, Cramer M, Plenio MB (2010) Colloquium: area laws for the entanglement entropy. Rev Mod Phys 82:277–306 https://doi.org/10.1103/RevModPhys.82.277
55. Serbyn M, Papić Z, Abanin DA Local conservation laws and the structure of the many-body localized states. Phys Rev Lett 111:127201. https://doi.org/10.1103/PhysRevLett.111.127201
56. Huse DA, Nandkishore R, Oganesyan V (2014) Phenomenology of fully many-body-localized systems. Phys Rev B 90:174202. https://doi.org/10.1103/PhysRevB.90.174202
57. Kagan Y, Maksimov LA (1984) Localization in a system of interacting particles diffusing in a regular crystal. Sov Phys JETP 60:201. 0038-5646/84/070201-10
58. Schiulaz M, Müller M (2014) Ideal quantum glass transitions: many-body localization without quenched disorder. In: AIP Conference Proceedings, vol. 1610, pp. 11–23. https://doi.org/10.1063/1.4893505
59. Schiulaz M, Silva A, Müller M (2015) Dynamics in many-body localized quantum systems without disorder. Phys Rev B 91:184202. https://doi.org/10.1103/PhysRevB.91.184202
60. Yao NY, Laumann CR, Cirac JI, Lukin MD, Moore JE (2016a) Quasi-many-body localization in translation-invariant systems. Phys Rev Lett 117:240601. https://doi.org/10.1103/PhysRevLett.117.240601
61. Mikheev AV, Maidano A, Mikhin NP (1983) Localization and quantum diffusion of $He^3$ atoms stimulated by phonons in $He^4$ crystals. Solid State Commun 48:361
62. Papić Z, Stoudenmire EM, Abanin DA (2015) Many-body localization in disorder-free systems: The importance of finite-size constraints. Ann Phys (N Y) 362:714–725. https://doi.org/10.1016/j.aop.2015.08.024
63. Garrahan JP (2018) Aspects of non-equilibrium in classical and quantum systems: slow relaxation and glasses, dynamical large deviations, quantum non-ergodicity, and open quantum dynamics. Phys A Stat Mech Appl 504:130–154. https://doi.org/10.1016/j.physa.2017.12.149
64. van Horssen M, Levi E, Garrahan JP (2015) Dynamics of many-body localization in a translation-invariant quantum glass model. Phys Rev B 92:100305. https://doi.org/10.1103/PhysRevB.92.100305
65. Hickey JM, Genway S, Garrahan JP (2016) Signatures of many-body localisation in a system without disorder and the relation to a glass transition. J Stat Mech Theory Exp 2016:054047
66. Garrahan JP, Newman MEJ (2000) Glassiness and constrained dynamics of a short-range nondisordered spin model. Phys Rev E 62:7670–7678. https://doi.org/10.1103/PhysRevE.62.7670
67. Castelnovo C, Chamon C (2012) Topological quantum glassiness. Philos Mag 92:1–3. https://doi.org/10.1080/14786435.2011.609152
68. Chamon C, Goerbig MO, Moessner R, Cugliandolo LF (2017) Topological aspects of condensed matter physics. In: Chamon C, Goerbig MO, Moessner R, Cugliandolo LF (eds) Oxford University Press, Oxford. https://global.oup.com/academic/product/topological-aspects-of-condensed-matter-physics-9780198785781?cc=gb&lang=en&
69. Peskin ME, Schroeder DV (1995) An introduction to quantum field theory. Westview Press
70. Wiese U-J (2013) Ultracold quantum gases and lattice systems: quantum simulation of lattice gauge theories. Ann Phys 525:777–796. https://doi.org/10.1002/andp.201300104

71. Zohar E, Cirac JI, Reznik B (2016) Quantum simulations of lattice gauge theories using ultracold atoms in optical lattices. Rep Prog Phys 79:014401. https://doi.org/10.1088/0034-4885/79/1/014401
72. Kühn S, Cirac JI, Bañuls M-C (2014) Quantum simulation of the Schwinger model: a study of feasibility. Phys Rev A 90:042305. https://doi.org/10.1103/PhysRevA.90.042305
73. Kogut J, Susskind L (1975) Hamiltonian formulation of Wilson's lattice gauge theories. Phys. Rev D 11:395–408. https://doi.org/10.1103/PhysRevD.11.395
74. Kitaev AY (2003) Fault-tolerant quantum computation by anyons. Ann Phys (N Y) 303:2–30. https://doi.org/10.1016/S0003-4916(02)00018-0
75. Feynman RP, Leighton RB, Sands M (2010) The Feynman lectures on physics. Volume II: Mainly electromagnetism and matter, New Millen edn. Basic Books, New York
76. Wegner FJ (1971) Duality in generalized ising models and phase transitions without local order parameters. J Math Phys 12:2259–2272. https://doi.org/10.1063/1.1665530
77. Kogut JB (1979) An introduction to lattice gauge theory and spin systems. Rev Mod Phys 51:659–713. https://doi.org/10.1103/RevModPhys.51.659
78. Anderson MH, Ensher JR, Matthews MR, Wieman CE, Cornell EA (1995) Observation of Bose-Einstein condensation in a dilute atomic vapor. Science 269:198–201. https://doi.org/10.1126/science.269.5221.198
79. Bakr WS, Gillen JI, Peng A, Fölling S, Greiner M (2009) A quantum gas microscope for detecting single atoms in a Hubbard-regime optical lattice. Nature 462:74–77. https://doi.org/10.1038/nature08482
80. Choi J-Y, Hild S, Zeiher J, Schauss P, Rubio-Abadal A, Yefsah T, Khemani V, Huse DA, Bloch I, Gross C (2016) Exploring the many-body localization transition in two dimensions. Science 352:1547–1552. https://doi.org/10.1126/science.aaf8834
81. Feynman RP (1982) Simulating physics with computers. Int J Theor Phys 21:467–488. https://doi.org/10.1007/BF02650179
82. Aspuru-Guzik A, Walther P (2012) Photonic quantum simulators. Nat Phys 8:285–291. https://doi.org/10.1038/nphys2253
83. Houck AA, Türeci HE, Koch J (2012) On-chip quantum simulation with superconducting circuits. Nat Phys 8:292–299. https://doi.org/10.1038/nphys2251
84. Weimer H, Müller M, Büchler HP, Lesanovsky I (2011) Digital quantum simulation with Rydberg atoms. Quantum Inf Process 10:885–906. https://doi.org/10.1007/s11128-011-0303-5
85. Kim H, Park Y, Kim K, Sim H-S, Ahn J (2018) Detailed balance of thermalization dynamics in Rydberg-atom quantum simulators. Phys Rev Lett 120:180502. https://doi.org/10.1103/PhysRevLett.120.180502
86. Cirac JI, Zoller P (1995) Quantum computations with cold trapped ions. Phys Rev Lett 74:4091–4094. http://arxi.org/abs/0305129 arXi:0305129 [quant-ph]. https://doi.org/10.1103/PhysRevLett.74.4091
87. Martinez EA, Muschik CA, Schindler P, Nigg D, Erhard A, Heyl M, Hauke P, Dalmonte M, Monz T, Zoller P, Blatt R (2016) Real-time dynamics of lattice gauge theories with a few-qubit quantum computer. Nature 534:516–519. https://doi.org/10.1038/nature18318
88. Bloch I, Dalibard J, Zwerger W (2008) Many-body physics with ultracold gases. Rev Mod Phys 80:885–964. https://doi.org/10.1103/RevModPhys.80.885
89. Aubry S, André G (1980) Analyticity breaking and anderson localization in incommensurate lattices. Ann Isreal Phys Soc 3:18
90. Bordia P, Lüschen H, Scherg S, Gopalakrishnan S, Knap M, Schneider U, Bloch I (2017) Probing slow relaxation and many-body localization in two-dimensional quasiperiodic systems. Phys Rev X 7:041047. https://doi.org/10.1103/PhysRevX.7.041047

# Chapter 2
# The Model

The main contribution of this thesis is the introduction of a family of models that have a disorder-free mechanism for localization. In this chapter we define the models and reveal an exact mapping to free fermions using a local $\mathbb{Z}_2$ gauge symmetry. This mapping unveils the mechanism for localization and allows us to perform efficient large-scale numerical simulations to demonstrate the localization behaviour, which we do in the subsequent chapters. In this chapter we will also make concrete connections between our Hamiltonians and the Hubbard model [1, 2], the Falicov-Kimball model [3] and Kitaev's toric code [4].

We study a family of lattice models with spinless fermions, $\hat{f}_i$, on the sites of a lattice coupled to spins-1/2, $\hat{\sigma}_{jk}$, positioned on the bonds. These models can be defined on an arbitrary graph. However, in this thesis we focus on one-dimensional chains and a two-dimensional square lattice with both open and periodic boundary conditions. We also discuss effects of perturbations in Chap. 6. The models are described by the Hamiltonian

$$\hat{H} = -J \sum_{\langle jk \rangle} \hat{\sigma}^z_{jk} \hat{f}^\dagger_j \hat{f}_k - h \sum_j \hat{A}_j, \tag{2.1}$$

where $\langle jk \rangle$ denotes nearest neighbours, and $\hat{A}_j$ is the star operator, which is the product of the spins on bonds connected to site $j$,

$$\hat{A}_j = \prod_{k \in +_j} \hat{\sigma}^x_{jk}, \tag{2.2}$$

shown for a 2D square lattice in Fig. 2.1. These star operators were introduced in Sect. 1.5.3 in the context of Kitaev's toric code [4]. They appear here with strength $h$, while $J$ defines the coupling between spins and fermions.


A. Smith, *Disorder-Free Localization*, Springer Theses,
https://doi.org/10.1007/978-3-030-20851-6_2

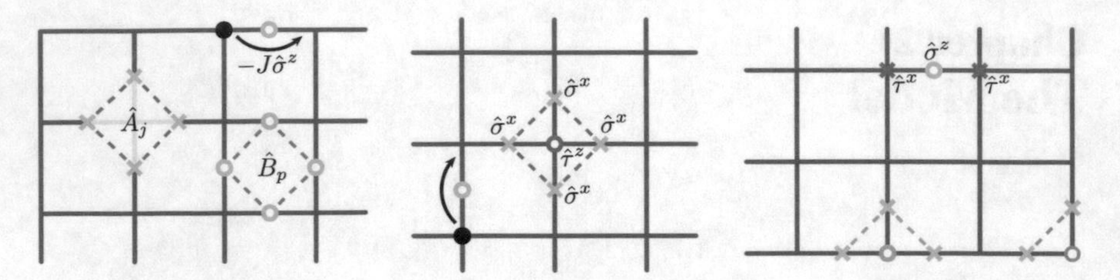


**Fig. 2.1** Schematic picture of the model (2.1). Left panel shows star $\hat{A}$ and plaquette $\hat{B}$ operators, with $\hat{\sigma}^x, \hat{\sigma}^z$ operators, denoted by crosses and circles, respectively. Fermion hopping has amplitude $J$ and a sign which depends on the $z$-component of the spin-1/2 on that bond. Centre and right panels show the duality transformation to new spins $\hat{\tau}$. The model can be defined with periodic boundary conditions, as in the centre panel. In the case of open boundaries, we define incomplete boundary "stars" as shown in the right panel. The figure is reproduced from the journal Physical Review B [5]

The Hamiltonian (2.1) possesses an extensive number of conserved quantities (charges) $\hat{q}_i = (-1)^{\hat{n}_i} \hat{A}_i$, where $\hat{n}_i = \hat{f}_i^\dagger \hat{f}_i$. These charges have eigenvalues $\pm 1$, they commute with the Hamiltonian and amongst themselves—i.e., $[\hat{H}, \hat{q}_i] = 0$, and $[\hat{q}_i, \hat{q}_j] = 0$—and they generate local $\mathbb{Z}_2$ gauge transformations under which the Hamiltonian is invariant. Explicitly, these transformations are implemented by the unitary operators $\hat{U}(\{\theta_i\}) = \prod_i \hat{q}_i^{(1-\theta_i)/2}$, where $\theta_i = \pm 1$, which transform the operators according to

$$\hat{f}_i \to \theta_i \hat{f}_i, \qquad \hat{\sigma}^z_{ij} \to \theta_i \theta_j \hat{\sigma}^z_{ij}. \tag{2.3}$$

It is worth noting that our model is an example of an unconstrained $\mathbb{Z}_2$ lattice gauge theory—that is, while the Hamiltonian (2.1) is invariant under the gauge transformation, the Hilbert space is not. What is typically understood as a gauge theory is constrained to the physical subspace of gauge invariant states by Gauss' law [1, 6, 7] $\hat{q}_i|\Psi\rangle = |\Psi\rangle$, which we do not impose in our case, cf. the gauge structure of the Kitaev honeycomb model [8].

In our investigation in later chapters we will predominantly consider the Hamiltonian (2.1) defined on a 1D chain and so for concreteness we will write down the specific form of the Hamiltonian. In this case the star operators reduce to nearest-neighbour Ising exchange couplings $\hat{A}_j = \hat{\sigma}^x_{j-1,j}\hat{\sigma}^x_{j,j+1}$, and the Hamiltonian (2.1) assumes the following form

$$\hat{H}_{\text{1D}} = -J \sum_{\langle jk \rangle} \hat{\sigma}^z_{jk} \hat{f}_j^\dagger \hat{f}_k - h \sum_j \hat{\sigma}^x_{j-1,j} \hat{\sigma}^x_{j,j+1}, \tag{2.4}$$

that is, a quantum Ising model coupled to spinless fermions.

## 2.1 Duality Mapping

The central step in our analysis of the Hamiltonian (2.1) is to perform a duality transformation for the spins. Through this mapping, we reveal an equivalence between charge configurations $\{q_i = \pm 1\}$ and configurations of on-site potentials for free fermions. Furthermore, we will show that the computations of observable quantities correspond to free fermion correlators averaged over binary disorder. In this Section, we explain this mapping in detail.

We proceed by a duality transformation of the operators $\sigma$, defining spin-1/2 operators $\tau$ that live on the sites of the lattice,

$$\hat{\tau}_j^z = \hat{A}_j, \qquad \hat{\tau}_j^x \hat{\tau}_k^x = \hat{\sigma}_{jk}^z, \tag{2.5}$$

as shown in Fig. 2.1, where the indices $j$ and $k$ correspond to nearest neighbour sites. This is the same kind of duality mapping as used in the context of the Ising gauge theory, see Sect. 1.5.2. As with any duality transformation, we have implicitly restricted ourselves to one of the disconnected subspaces of the model. These disconnected subspaces can be enumerated by another set of conserved quantities defined as products of $\hat{\sigma}^z$ along closed loops on the lattice. These conserved quantities can be expressed in terms of plaquette operators, $\hat{B}_p$, defined on the irreducible plaquettes of the lattice (see Fig. 2.1) and Wilson loop operators $\hat{\Gamma}_n$,

$$\hat{B}_p = \prod_{\text{plaquette } p} \hat{\sigma}_{jk}^z, \qquad \hat{\Gamma}_n = \prod_{\langle jk \rangle \in \gamma_n} \hat{\sigma}_{jk}^z, \tag{2.6}$$

where $\gamma_n$ is any non-contractible closed path such that $\hat{\Gamma}_n$ cannot be written as a product of plaquette operators. For concreteness, we give two examples: in a 1D periodic chain there is only one such operator which is the loop around the entire system: $\prod_{\text{all}} \hat{\sigma}^z$. This is equivalent to a statement that the number of domain walls modulo 2 is conserved. On a 2D periodic square lattice we have $\hat{B}_p$ defined on all the square plaquettes and two Wilson loops around the two periodic directions. Importantly, as well as commuting with the Hamiltonian, these operators commute with the generators of the $\mathbb{Z}_2$ gauge symmetry $\hat{q}_i$. The eigenvalues $\pm 1$ of these operators label subspaces which are disconnected under gauge transformations. Our choice of duality mapping (2.5) is only valid in the sector with all plaquette operators $\hat{B}_p = 1$ and all loop operators $\hat{\Gamma}_n = 1$. Other sectors can be accessed using different duality transformations [9].

In terms of the $\tau$ spins, the Hamiltonian assumes the form

$$\hat{H} = -J \sum_{\langle jk \rangle} \hat{\tau}_j^x \hat{\tau}_k^x \hat{f}_j^\dagger \hat{f}_k - h \sum_j \hat{\tau}_j^z. \tag{2.7}$$

Although this Hamiltonian is equivalent to Eq. (2.1) only on a restricted Hilbert space, we will not use notation to differentiate between the two. In this form we can identify the local conserved charges as $\hat{q}_j = \hat{\tau}_j^z(-1)^{\hat{n}_j}$ with $\hat{n}_j = \hat{f}_j^\dagger \hat{f}_j$. The charges are precisely those that generate the gauge symmetry identified in the original degrees of freedom.

Finally, by a change of variables $\hat{c}_j = \hat{\tau}_j^x \hat{f}_j$, the Hamiltonian (2.7) can be written in terms of the conserved charges, $\hat{q}_j$, and the spinless fermions, $\hat{c}_j$:

$$\hat{H} = -J \sum_{\langle jk \rangle} \hat{c}_j^\dagger \hat{c}_k + 2h \sum_j \hat{q}_j (\hat{c}_j^\dagger \hat{c}_j - 1/2), \tag{2.8}$$

where we have used the fact that $\hat{n}_j = \hat{f}_j^\dagger \hat{f}_j = \hat{c}_j^\dagger \hat{c}_j$, since $(\hat{\tau}_j^x)^2 = 1$. The canonical anti-commutation relations $\{\hat{c}_j^\dagger, \hat{c}_k\} = \delta_{jk}$ for these new spinless fermions can be similarly verified. For a given charge configuration—that is, in the subspace of fixed $\{q_j\} = \pm 1$—the Hamiltonian (2.8) describes a fermion tight-binding model with a binary potential whose sign at each site is given by the value of $q_j$. In a randomly selected charge sector this corresponds to the Anderson model of localization in Eq. (1.21), but with the potential selected at random from $\{-2h, 2h\}$, i.e., from a bimodal distribution.

### 2.1.1 Global Constraints

In defining the duality transformation (2.5) we also have a global constraint on the $\tau$ spins. In all but one dimensions with open boundary conditions we have that the product of all star operators is the identity, i.e.,

$$\prod_{\text{all } j} \hat{A}_j = \mathbb{1}. \tag{2.9}$$

This is because each Pauli matrix $\hat{\sigma}_{jk}^x$ appears in exactly two star operators and so they appear in this product exactly twice and square to the identity. While this is a trivial consequence of the form of the star operators in terms of the $\sigma$ spins, this imposes a non-trivial constraint on the $\tau$ spins

$$\prod_{\text{all } j} \hat{\tau}_j^z = \mathbb{1}. \tag{2.10}$$

This constraint is not imposed with open boundary conditions in one dimension. In this case the $\tau$ spins on the end sites do not enter the Hamiltonian, and so without loss of generality we will impose this constraint in all cases. Equations (2.9) and (2.10) have the consequence that the product of all charges is determined by the total fermion parity

$$\prod_{\text{all } j} \hat{q}_j = (-1)^{\hat{N}_f}, \tag{2.11}$$

where $\hat{N}_f = \sum_j \hat{n}_j$ is the total fermion number.

### 2.1.2 Transformation of States

As well as understanding how the operators transform under the mapping, we must also make an identification between states. Let us consider a basis of tensor product states of the form $|\Psi\rangle_{\sigma,f} = |S\rangle_\sigma \otimes |\psi\rangle_f$, which we wish to identify with a state $|\Psi\rangle_{\tau,c}$ in the Hilbert space of the $\tau$ and $c$ degrees of freedom, and then in turn with $|\Psi\rangle_{q,c}$. If for the fermion states we choose the Fock states, i.e., $|\psi\rangle_f = \hat{f}_j^\dagger \cdots \hat{f}_l^\dagger |vacuum\rangle$, then since $\hat{c}_i^\dagger \hat{c}_i = \hat{f}_i^\dagger \hat{f}_i$ these states take the same form for the $c$ fermions, and we will drop the subscript in the following. Without loss of generality, let us consider spins in the $z$-polarized state $|\uparrow\uparrow\uparrow\cdots\rangle_\sigma$—any other spin state in the sector defined by all $\hat{B}_j = 1$ can be reached via application of star operators $\hat{A}_j$. By the duality transformation (2.5) and the constraint in Eq. (2.10), this initial state satisfies

$$\hat{\tau}_j^x \hat{\tau}_k^x |\uparrow\uparrow\uparrow\cdots\rangle_\sigma = |\uparrow\uparrow\uparrow\cdots\rangle_\sigma, \qquad \prod_{\text{all} j} \hat{\tau}_j^z |\uparrow\uparrow\uparrow\cdots\rangle_\sigma = |\uparrow\uparrow\uparrow\cdots\rangle_\sigma, \tag{2.12}$$

where sites $j$ and $k$ are nearest-neighbours. The left set of conditions implies the form $|\uparrow\uparrow\uparrow\cdots\rangle_\sigma = \alpha|\rightarrow\rightarrow\cdots\rangle_\tau + \beta|\leftarrow\leftarrow\cdots\rangle_\tau$, and with the remaining constraint we make the correspondence

$$|\uparrow\uparrow\uparrow\cdots\rangle_\sigma = \frac{1}{\sqrt{2}}\big(|\rightarrow\rightarrow\cdots\rangle_\tau + |\leftarrow\leftarrow\cdots\rangle_\tau\big). \tag{2.13}$$

We can generally make identifications of the form $|S\rangle_\sigma \otimes |\psi\rangle \propto |S\rangle_\tau \otimes |\psi\rangle$.

Finally, we can express these tensor product states in terms of conserved charges instead of the $\tau$ spins. For a $z$-polarized state this identification proceeds as follows

$$|\uparrow\uparrow\uparrow\cdots\rangle_\sigma \otimes |\psi\rangle = \frac{1}{\sqrt{2^{N-1}}} \sum_{\{\tau_i\}=\uparrow,\downarrow}' |\tau_1, \tau_2 \cdots\rangle_\tau \otimes |\psi\rangle, \tag{2.14}$$

where we have identified $|\rightarrow\rangle_\tau = (|\uparrow\rangle_\tau + |\downarrow\rangle_\tau)/\sqrt{2}$, for each $\tau$ spin. The prime indicates that the sum is over all possible spin configurations that satisfy the constraint in Eq. (2.10), which is reflected in the normalization coefficient. Let us consider a single state in this sum, $|\tau_1\tau_2\cdots\rangle_\tau \otimes |\psi\rangle$, then since the fermion state is a simple tensor product of site occupation, this can be rewritten as

$$|\tau_1(-1)^{n_1}, \tau_2(-1)^{n_2}, \cdots\rangle_q \otimes |\psi\rangle. \tag{2.15}$$

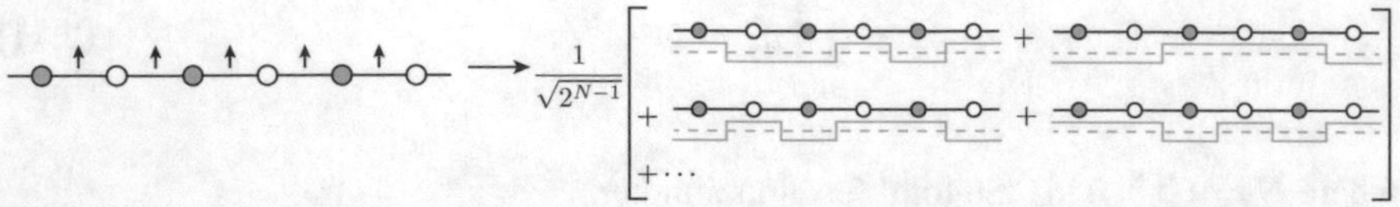

Fig. 2.2 Schematic picture showing the transformation of the initial state into a dual representation. On the left is an initial state with fermions in a charge density wave (filled sites are blue, empty sites are white). The bond spins are polarized along the $z$-axis. The dual state (right panel) has the fermions in the same configuration, but the wave-function is an equal superposition of charge configurations. Each charge configuration corresponds to a different binary potential for the fermions, shown in grey. The figure is reproduced from the journal Physical Review B [5] (color figure online)

The occupation numbers for the fermion state are fixed and thus only contribute a common sign structure to the charge configuration. Since we sum over all $\tau$ configurations in Eq. (2.14) with a positive weight, this equates to an equal sum over charge configurations

$$|\uparrow\uparrow\cdots\rangle_\sigma \otimes |\psi\rangle = \frac{1}{\sqrt{2^{N-1}}} \sum_{\{q_j\}=\pm 1}' |q_1, q_2, \cdots, q_N\rangle \otimes |\psi\rangle, \tag{2.16}$$

where again the prime indicates that the sum is over all charge configurations that satisfy the constraint in Eq. (2.11). This transformation of states is shown schematically in Fig. 2.2. The fact that all of the weights are equal and positive is important for this final form, otherwise there would be a sign structure that depends both on the spin and on the fermion configuration. Other spin states in the same spin sector can be accessed through the application of star operators. Except briefly in Sect. 2.4, we will only consider initial states of the form $|\Psi\rangle = |\uparrow\uparrow\uparrow\cdots\rangle_\sigma \otimes |\psi\rangle$ in this thesis.

## 2.1.3 *Explicit Example in 1D*

To aid the above discussion of the transformation of states, we will consider the concrete example of a 1D chain with periodic boundary conditions in the initial state $|\Psi\rangle = |\uparrow\uparrow\uparrow\rangle_\sigma \otimes |101\rangle$. The fermion state has the same form in terms of the $c$ fermions, but the spin state transforms as

$$|\uparrow\uparrow\uparrow\rangle_\sigma = \frac{1}{\sqrt{2}}\big(|\rightarrow\rightarrow\rightarrow\rangle_\tau + |\leftarrow\leftarrow\leftarrow\rangle_\tau\big). \tag{2.17}$$

If we change from the $x$ to the $z$ basis for the $\tau$-spins then the transformation reads

$$|\Psi\rangle = \frac{1}{2}\big(|\uparrow\uparrow\uparrow\rangle_\tau + |\uparrow\downarrow\downarrow\rangle_\tau + |\downarrow\uparrow\downarrow\rangle_\tau + |\downarrow\downarrow\uparrow\rangle_\tau\big) \otimes |101\rangle, \tag{2.18}$$

which can be seen to match Eq. (2.14) with a sum over all $\tau$-states satisfying the condition $\prod_{\text{all } j} \hat{\tau}_j^z = \mathbb{1}$. Finally, we can transform to the representation in terms of conserved charges, which gives

$$|\Psi\rangle = \frac{1}{2}\big(|-+-\rangle_q + |--+\rangle_q + |+++\rangle_q + |+--\rangle_q\big) \otimes |101\rangle, \tag{2.19}$$

where we have written all terms in the same order as in Eq. (2.18). Again, we see that the sum includes all terms that satisfy the constraint on the charges with equal weight, as in Eq. (2.16).

Let us consider an example of another initial spin state that can be accessed through the application of the star operators $\hat{A}_j$. While we will not consider such states hereafter in this thesis, this example illustrates the sign structure that such states have relative to the $z$-polarized state. In 1D, the star operators are of the form $\hat{\sigma}_j^x \hat{\sigma}_{j+1}^x$. If we act on the initial state with $\hat{A}_2$ then this results in the spin state $\hat{A}_2|\uparrow\uparrow\uparrow\rangle_\sigma = |\downarrow\downarrow\uparrow\rangle_\sigma$. In terms of $\tau$ spins the star operators are the $\hat{\tau}_j^z$ operators and we therefore have

$$\begin{aligned} \hat{A}_j|\Psi\rangle &= \frac{1}{2}\big(|\uparrow\uparrow\uparrow\rangle_\tau - |\uparrow\downarrow\downarrow\rangle_\tau + |\downarrow\uparrow\downarrow\rangle_\tau - |\downarrow\downarrow\uparrow\rangle_\tau\big) \otimes |101\rangle \\ &= \frac{1}{2}\big(|-+-\rangle_q - |--+\rangle_q + |+++\rangle_q - |+--\rangle_q\big) \otimes |101\rangle, \end{aligned} \tag{2.20}$$

where we note the additional sign structure of the state. Everything in this thesis can be repeated using different initial spin states, but it is crucial that this sign structure must be taken into account.

## 2.2 Emergent Disorder and Disorder Averaging

In the previous section we gave the details of a transformation of our model (2.1) to the Hamiltonian (2.8), which has an effective binary potential determined by the conserved charges. Furthermore, we showed that the states in the dual language are superpositions of charge configurations. In order to make a connection with the Anderson localization problem, the final step is to show that expectation values of observables with respect to these initial states amount to averages over the effective disorder.

Let us, for concreteness, consider the spin expectation value

$$\langle\Psi|\hat{\sigma}_{jk}^z(t)|\Psi\rangle = \frac{1}{2^{N-1}} \sum_{\{s_l\},\{q_m\}=\pm 1}' \langle\psi| \otimes \langle s_1, \ldots| e^{i\hat{H}t} \hat{\tau}_j^x \hat{\tau}_k^x e^{-i\hat{H}t} |q_1, \ldots\rangle \otimes |\psi\rangle. \tag{2.21}$$

In order to simplify expressions, we introduce the Hamiltonian

$$\hat{H}(q) = -J \sum_{\langle jk \rangle} \hat{c}_j^\dagger \hat{c}_k + 2h \sum_j q_j (\hat{c}_j^\dagger \hat{c}_j - 1/2), \tag{2.22}$$

where we use the shorthand notation $q = \{q_j\}$, which corresponds to a particular charge configuration. The difference between this Hamiltonian and that appearing in Eq. (2.8) is that the $q_i$ here are classical variables corresponding to a fixed charge configuration $\{q_j\} = \pm 1$, and it acts only on the fermion subspace. In Eq. (2.21) we then commute the $\tau$ operators past the unitary time evolution using the fact that $\hat{\tau}_j^x \hat{H}(q) = \hat{H}_j(q)\hat{\tau}_j^x$, where the subscript $j$ signifies that the value of charge $q_j$ in the Hamiltonian has been reversed relative to $\hat{H}(q)$. The operator $\hat{\tau}_j^x \hat{\tau}_k^x$ then acts trivially on the initial spin state, see Eq. (2.12). The spin expectation value can then be written as

$$\langle \Psi | \hat{\sigma}_{jk}^z(t) | \Psi \rangle = \frac{1}{2^{N-1}} \sum_{\{q_i\}=\pm 1}' \langle \psi | e^{i\hat{H}(q)t} e^{-i\hat{H}_{jk}(q)t} | \psi \rangle, \tag{2.23}$$

where we have removed the second sum over charges using charge conservation. This is in the form of a free-fermion correlator averaged over charge configurations, which amount to configurations of the potential in the Hamiltonian (2.22), see also Ref. [10]. The correlators can be efficiently computed using determinants, see Appendix B. It is important to note that, as in Eq. (2.23), the expressions for the correlators that we obtain are generally distinct from, e.g., fermion correlators of a tight binding model with disorder. For instance, the Green's function

$$\langle \hat{f}_j^\dagger(t) \hat{f}_k(0) \rangle = \frac{1}{2^{N-1}} \sum_{\{q_i\}=\pm 1}' \langle \psi | e^{i\hat{H}(q)t} \hat{c}_j^\dagger e^{-i\hat{H}_j(q)t} \hat{c}_k | \psi \rangle \tag{2.24}$$

does not correspond to averaging over disorder configurations for the Green's functions $\langle \hat{c}_j^\dagger(t) \hat{c}_k(0) \rangle$ because of the flipped charges between the forward and backward time evolution. In this respect, the correlators that we obtain are similar to the ones appearing in the X-ray edge problem [11] and the dynamical structure factor for the honeycomb Kitaev model [12, 13], which correspond to local quantum quenches.

## 2.3 Connections to Other Models

Our model is closely connected to range of important theoretical models and here we will briefly outline a few important examples. For further discussion we point the reader to the literature.

### 2.3.1 The Hubbard Model

The first example that we make connection with is the Hubbard model [1, 2, 14], which is the paradigmatic model of strongly correlated electrons. It is a simplified model of electrons in a tight-binding potential of a solid, interacting via coulomb interactions.

The Hubbard model is constructed by using the localized Wannier states

$$\Psi_{\mathbf{r}_i}(\mathbf{r}) = \frac{1}{\sqrt{N}} \sum_{\mathbf{k} \in \text{B.Z.}} e^{i\mathbf{k}\cdot\mathbf{r}_i} \Psi_{\mathbf{k}}(\mathbf{r}), \tag{2.25}$$

where $\Psi_{\mathbf{k}}(\mathbf{r})$ are the single particle eigenstates of the Hamiltonian $\hat{H} = \mathbf{p}^2/2m + V(\mathbf{r})$, where $V(\mathbf{r})$ is the potential felt by the electrons due to the structure of the crystal lattice. These eigenstates are the Bloch wavefunctions and take the form $\Psi_{\mathbf{k}}(\mathbf{r}) = e^{i\mathbf{k}\cdot\mathbf{r}} u_{\mathbf{k}}(\mathbf{r})$, where $\mathbf{k}$ is the quasi-momentum and $u_{\mathbf{k}}(\mathbf{r})$ has the same periodicity as the lattice. The Wannier states are localized at the lattice sites $\mathbf{r}_i$. Here we consider a single band and orbital, but the extension to multiple orbitals and bands is straightforward. The electrons interact via

$$U_{ijkl} = \int \mathrm{d}\mathbf{r}_1 \mathrm{d}\mathbf{r}_2 \; \Psi^*_{\mathbf{r}_i}(\mathbf{r}_1) \Psi^*_{\mathbf{r}_j}(\mathbf{r}_2) V(\mathbf{r}_1 - \mathbf{r}_2) \Psi_{\mathbf{r}_k}(\mathbf{r}_2) \Psi_{\mathbf{r}_l}(\mathbf{r}_1), \tag{2.26}$$

where $V$ is the screened Coulomb interaction. We then make a tight-binding approximation [14] keeping only nearest-neighbour hopping, and we keep only the dominant on-site interactions ($i = j = k = l$). The result is the Hubbard model, described by the Hamiltonian

$$\hat{H} = - \sum_{\langle i,j \rangle,\, \alpha=\uparrow,\downarrow} t_{ij} \left( \hat{c}^\dagger_{i,\alpha} \hat{c}_{j,\alpha} + \text{H.c.} \right) + U \sum_i \hat{n}_{i,\uparrow} \hat{n}_{i,\downarrow}, \tag{2.27}$$

written in second quantised notation, where $\hat{c}^\dagger_{i,\alpha}$ creates a fermion in the Wannier state $\Psi_{\mathbf{r}_i}$ with spin $\alpha$, and $\hat{n}_{i,\alpha} = \hat{c}^\dagger_{i,\alpha} \hat{c}_{i,\alpha}$.

Let us now consider a generalized version of our model in 1D, namely

$$\hat{H} = -J \sum_{j,\, \alpha=\uparrow,\downarrow} \hat{\sigma}^z_{j,j+1} \left( \hat{f}^\dagger_{j,\alpha} \hat{f}_{j+1,\alpha} + \text{H.c.} \right) - h \sum_j \hat{\sigma}^x_{j-1,j} \hat{\sigma}^x_{j,j+1}, \tag{2.28}$$

where we have added a spin degree of freedom to the fermions and have the same coupling to the $\sigma$-spins for both species of fermions. This model also possesses an extensive set of conserved quantities, $\hat{q}_j = \hat{\sigma}^x_{j-1,j} \hat{\sigma}^x_{j,j+1} (-1)^{\hat{n}_{j,\uparrow} + \hat{n}_{j,\downarrow}}$, and we can perform a duality transformation of the spins as before—that is, we define

$$\hat{\tau}^z_j = \hat{\sigma}^x_{j-1,j} \hat{\sigma}^x_{j,j+1}, \qquad \hat{\tau}^x_j, \hat{\tau}^x_{j+1} = \hat{\sigma}^z_{j,j+1}, \qquad \hat{c}_{j,\alpha} = \hat{\tau}^x_j \hat{f}_{j,\alpha}. \tag{2.29}$$

In this dual language, the conserved quantities take the form $\hat{q}_j = \hat{\tau}_j^z (-1)^{\hat{n}_{j,\uparrow}+\hat{n}_{j,\downarrow}}$ and the Hamiltonian reads

$$\hat{H} = -J \sum_{j,\alpha} \left( \hat{c}_{j,\alpha}^\dagger \hat{c}_{j+1,\alpha} + \text{H.c.} \right) - 4h \sum_j \hat{q}_j \hat{n}_{j,\uparrow} \hat{n}_{j,\downarrow} + 2h \sum_j \hat{q}_j (\hat{n}_{j,\uparrow} + \hat{n}_{j,\downarrow}) - h \sum_j \hat{q}_j. \tag{2.30}$$

The first two terms are those appearing in the Hubbard model but with site-dependent interactions set by the charge configuration $\{q_j\} = \pm 1$. The next two terms are a $q$-dependent on-site potential and an energy shift which depends on the charge configuration. Unfortunately, this model does not have a free fermion limit and we leave the study of its dynamics for future work.

This model has been considered in Refs. [15, 16] as a slave-spin description of the Hubbard model, which is recovered by restricting the Hilbert space to those states satisfying $\hat{q}_j |\psi\rangle = |\psi\rangle$, i.e., $\hat{\tau}_j^z (-1)^{\hat{n}_{j,\uparrow}+\hat{n}_{j,\downarrow}} |\psi\rangle = |\psi\rangle$. References [15, 16] use mean field theory on the larger unconstrained Hilbert space as a way to approach the Hubbard model.

### 2.3.2 Falicov-Kimball Model

The next model that we will make a connection with is the Falicov-Kimball model [3, 17–19]. This is an asymmetric version of the Hubbard model where one of the two species of fermions is frozen and cannot hop. The model is usually written as

$$\hat{H} = -J \sum_j \left( \hat{f}_j^\dagger \hat{f}_{j+1} + \text{H.c.} \right) + U \sum_j \hat{f}_j^\dagger \hat{f}_j \, \hat{g}_j^\dagger \hat{g}_j, \tag{2.31}$$

where $\hat{f}$ and $\hat{g}$ are spinless fermion operators which replace the two spin species of fermion, $\hat{f}_\uparrow$ and $\hat{f}_\downarrow$, from the Hubbard model. The $g$-fermions are static and provide an effective on-site potential for the $f$-fermions. The model is usually studied with fixed filling for both species of fermions.

In connection to our model we can define the charges $\hat{q}_i = 2\hat{g}_i^\dagger \hat{g}_i - 1$, which commute with the Hamiltonian (2.31). The Hamiltonian then takes exactly the same form as Eq. (2.8). The important difference is that the Falicov-Kimball model is defined with fixed filling, which for the charges would take the form $\sum_i (\hat{q}_i + 1)|\Psi\rangle = 2\mu|\Psi\rangle$, where $\mu$ is an integer that corresponds to the chemical potential for the $g$-fermions. We do not impose this in our model and we consider all charge sectors.

### 2.3.3 The Disorder-Free Localization Mechanism in Other Models

Since the publication of our work on disorder-free localization in Refs. [20, 21], there have been several other works that study the same or a similar localization mechanism in different settings.

The most direct extension is found in work by Prosko et al. [9] where they consider the Hamiltonian (2.1) in one and two dimensions with spinless fermions, spinfull fermions, and Majorana fermions. Furthermore, they consider the models both with Gauss' law imposed and also unconstrained. Rather than dynamics, they are interested in the ground state phase diagrams of such models.

In Ref. [22], Brenes et al. study the unconstrained lattice Schwinger model which has a local $U(1)$ gauge symmetry, rather than $\mathbb{Z}_2$, as in our case. They also show that the associated conserved charges can play the role of effective disorder for the fermions. In that setting, there is no free-fermion limit and instead they demonstrate MBL behaviour using numerical simulations based on Krylov subspace methods, see Appendix C.

## 2.4 Defect Attachment in the Toric Code Model

Before moving on to study the physics of localization, we would first like to mention a connection to Kitaev's toric code [4]. The Hamiltonian (2.1), defined on a 2D square lattice with periodic boundary conditions, is equivalent to the toric code (with plaquette dynamics frozen) coupled to spinless fermions. This coupling to the fermions induces dynamics for the star defects, which we will briefly discuss here. Note that this coupling is similar to a transverse magnetic field, but here it is dependent on the hopping of the fermions.

For our discussion of localization in the remainder of this thesis, we study initial states with spins polarized along the $z$-axis. However, let us for the moment focus on the spins in the ground state $|S_0\rangle$ of the toric code, that is

$$\hat{A}_j|S_0\rangle = |S_0\rangle, \qquad \hat{B}_p|S_0\rangle = |S_0\rangle. \tag{2.32}$$

We note that our choice of duality transformation is consistent with this ground state, and it also fixes the Wilson loop operators (2.6) to be $\hat{\Gamma}_1 = \hat{\Gamma}_2 = 1$, and so the duality transformation uniquely chooses one of the four degenerate ground states of the toric code. To access other ground state sectors we can modify the duality transformation (2.5) by defining a vertical and a horizontal line (going through bonds) across which the implicit definition of $\hat{\tau}^x$, picks up a sign $\Gamma_1, \Gamma_2 = \pm 1$, respectively. More explicitly, we can define

$$\hat{\tau}^x_j \hat{\tau}^x_k = (\Gamma_1)^{\delta^1_{jk}} (\Gamma_2)^{\delta^2_{jk}} \hat{\sigma}^z_{jk}, \tag{2.33}$$

where $\delta_{jk}^{1(2)}$ is 1 when the bond $\langle jk \rangle$ crosses a vertical (horizontal) reference line, and 0 otherwise. Note that this choice changes the action of the Wilson loop operators, but not $\hat{B}_p$, since any plaquette crosses any line an even number of times.

Having chosen our initial spin configuration we can consider the coupling to fermions. For a simple tensor product state, $|S_0\rangle \otimes |\psi\rangle$, this maps to

$$|S_0\rangle \otimes |\psi\rangle = |\uparrow\uparrow\cdots\rangle_\tau \otimes |\psi\rangle = |(-1)^{\hat{n}_1}, (-1)^{\hat{n}_2}, \cdots\rangle_q \otimes |\psi\rangle, \quad (2.34)$$

that is, for an initial fermion state of definite local occupation, the charge configuration is uniquely specified by the parities of fermion occupation numbers on each site. The Hamiltonian then takes a simple form

$$\hat{H}_{\text{TC}} = -J \sum_{\langle jk \rangle} \hat{c}_j^\dagger \hat{c}_k + 2h \sum_j q_j (\hat{c}_j^\dagger \hat{c}_j - 1/2), \quad (2.35)$$

where in contrast to Eq. (2.8), the charges $q_j$ are equal to $-1$ if there is a fermion on that site in the initial state, and $+1$ if the site is empty. If we consider the limit $h \gg J$, then the fermions lie at the bottom of large potential wells, fermion hopping is suppressed, and we recover the static toric code. The form of the conserved charges is a statement that defects of the toric code are attached to fermions (or holes).

Let us now consider excitations of this model in the limit of $h \gg J$. We can create a pair of defects at sites $j$ and $k$ by flipping the $\tau$ spins on those sites—the creation of a single defect is forbidden by the constraint in Eq. (2.10). These defects correspond to changing the sign of the potential on sites $j$ and $k$. These defects are then free to move in a restricted geometry on the lattice which has the opposite fermion parity to the fermion/hole attached to the defect. That is, the fermion attached to the defect can hop only between sites that have the same sign of the potential. This is a site percolation problem for the defects, and we will encounter it again in Sect. 3.4. Importantly, on a square lattice, the percolation threshold is $p_c \approx 0.5927$, which means that for fermions at half filling and in a random configuration, the defects are localized. There is then the possibility of interesting defect dynamics in such a model which depends crucially on the fermion filling but we leave a full investigation for future work.

As a final remark, let us return to the discussion of the plaquette operators, which can also be included in the Hamiltonian (2.1). Further, if we consider a dual lattice which is the square lattice with sites at the centres of plaquettes, then with respect to this lattice the plaquette operators become star operators and vice versa. We can then add a second fermion species on this dual lattice which will be attached to plaquette defects. If we denote fermions attached to star defects by $\hat{a}$ and those attached to plaquette defects by $\hat{b}$, then the corresponding Hamiltonian reads

$$\hat{H}_{\text{TC}} = -h_A \sum_s \hat{A}_s - h_B \sum_p \hat{B}_p - J_A \sum_{\langle ij \rangle_s} \hat{\sigma}_{j,k}^z \hat{a}_i^\dagger \hat{a}_j - J_B \sum_{\langle ij \rangle_p} \hat{\sigma}_{j,k}^x \hat{b}_i^\dagger \hat{b}_j, \quad (2.36)$$

where $\langle ij \rangle_s$ denotes nearest neighbours on the original lattice, and $\langle ij \rangle_p$ denotes those on the dual lattice. This model has the conserved charges $\hat{q}_j^A = \hat{A}_j(-1)^{\hat{a}_j^\dagger \hat{a}_j}$ and $\hat{q}_j^B = \hat{B}_j(-1)^{\hat{b}_j^\dagger \hat{b}_j}$, respectively. We now have the full toric code Hamiltonian with both types of defects attached to spinless fermions.

## References


1. Fradkin E (2013) Field theories of condensed matter physics. Cambridge University Press, Cambridge, UK. https://doi.org/10.1017/CBO9781139015509
2. Essler FHL, Frahm H, Göhman F, Klümper A, Korepin VE (2005) The one-dimensional Hubbard model. Cambridge University Press, Cambridge, UK
3. Falicov LM, Kimball JC (1969) Simple model for semiconductor-metal transitions: SmB6 and transition-metal oxides. Phys Rev Lett 22:997–999. https://doi.org/10.1103/PhysRevLett.22.997
4. Kitaev AY (2003) Fault-tolerant quantum computation by anyons. Ann Phys (NY) 303:2–30. https://doi.org/10.1016/S0003-4916(02)00018-0
5. Smith A, Knolle J, Moessner R, Kovrizhin DL (2018a) Dynamical localization in $Z_2$ lattice gauge theories. Phys Rev B 97:245137. https://doi.org/10.1103/PhysRevB.97.245137
6. Kogut JB (1979) An introduction to lattice gauge theory and spin systems. Rev Mod Phys 51:659–713. https://doi.org/10.1103/RevModPhys.51.659
7. Wen X-G (2007) Quantum field theory of many-body systems. Oxford University Press, Oxford, UK. https://doi.org/10.1093/acprof:oso/9780199227259.001.0001
8. Kitaev AY (2006) Anyons in an exactly solved model and beyond. Ann Phys (NY) 321:2–111. https://doi.org/10.1016/j.aop.2005.10.005
9. Prosko C, Lee S-P, Maciejko J (2017) Simple $Z_2$ gauge theories at finite fermion density. Phys Rev B 96:205104. https://doi.org/10.1103/PhysRevB.96.205104
10. Paredes B, Verstraete F, Cirac JI (2005) Exploiting quantum parallelism to simulate quantum random many-body systems. Phys Rev Lett 95:140501. https://doi.org/10.1103/PhysRevLett.95.140501
11. Gogolin AO, Nersesyan AA, Tsvelik AM (1998) Bosonization and strongly correlated systems. Cambridge University Press, Cambridge, UK
12. Baskaran G, Mandal S, Shankar R (2007) Exact results for spin dynamics and fractionalization in the Kitaev model. Phys Rev Lett 98:247201. https://doi.org/10.1103/PhysRevLett.98.247201
13. Knolle J, Kovrizhin DL, Chalker JT, Moessner R (2014) Dynamics of a two-dimensional quantum spin liquid: signatures of emergent Majorana fermions and fluxes. Phys Rev Lett 112:207203. http://journals.aps.org/prl/supplemental/10.1103/PhysRevLett.112.207203
14. Ashcroft NW, Mermin ND (1976) Solid state physics. Cengage Learning
15. Rüegg A, Huber SD, Sigrist M (2010) Z2-slave-spin theory for strongly correlated fermions. Phys Rev B 81:155118. https://doi.org/10.1103/PhysRevB.81.155118
16. Žitko R, Fabrizio M (2015) Z2 gauge theory description of the Mott transition in infinite dimensions. Phys Rev B 91:245130. https://doi.org/10.1103/PhysRevB.91.245130
17. Antipov AE, Javanmard Y, Ribeiro P, Kirchner S (2016) Interaction-tuned Anderson versus Mott localization. Phys Rev Lett 117:146601. https://doi.org/10.1103/PhysRevLett.117.146601
18. Herrmann AJ, Antipov AE, Werner P (2018) Spreading of correlations in the Falicov-Kimball model. Phys Rev B 97:165107. https://doi.org/10.1103/PhysRevB.97.165107
19. Gazit S, Randeria M, Vishwanath A (2017) Emergent Dirac fermions and broken symmetries in confined and deconfined phases of Z2 gauge theories. Nat Phys 13:484–490. https://doi.org/10.1038/nphys4028

20. Smith A, Knolle J, Kovrizhin DL, Moessner R (2017a) Disorder-free localization. Phys Rev Lett 118:266601. https://doi.org/10.1103/PhysRevLett.118.266601
21. Smith A, Knolle J, Moessner R, Kovrizhin DL (2017b) Absence of ergodicity without quenched disorder: from quantum disentangled liquids to many-body localization. Phys Rev Lett 119:176601. https://doi.org/10.1103/PhysRevLett.119.176601
22. Brenes M, Dalmonte M, Heyl M, Scardicchio A (2018) Many-body localization dynamics from gauge invariance. Phys Rev Lett 120:030601. https://doi.org/10.1103/PhysRevLett.120.030601

# Chapter 3
# Localization

In this chapter we investigate the localization behaviour in our model in one and two dimensions. The localization of the fermions is diagnosed using global quantum quench protocols relevant to experiments. Spectral properties of the model also shed light on transient behaviour observed in these dynamical probes that is unique to binary disorder. We pay particularly close attention to the strong effective disorder limit where we are able to perform a perturbation expansion, which in 2D allows us to make connection with a quantum site percolation problem. The latter provides a potential mechanism for a delocalization transition in two dimensions and higher.

Since the model can be mapped to free fermions we can calculate correlators using determinants as explained in Appendix B. This approach allows us to study large systems with $\sim 10^2$–$10^3$ sites. To calculate the density of states we use the kernel polynomial method [1], see Appendix D, which can be used for systems of order $10^5$–$10^6$ sites. Localization lengths are computed using a standard transfer matrix approach [2] described in Appendix E. All correlators studied in this section are found to be self-averaging and we approximate them by only averaging over a random selection of $\sim 10^3$–$10^4$ charge configurations.

## 3.1 Localization in 1D

We will first consider localization in the one dimensional version of our model (2.4), and will study the localization behaviour using a global quantum quench protocol. The initial states we consider have the bond-spins polarized along the $z$-axis $|\uparrow\uparrow\uparrow\cdots\rangle$, and fermions in one of the following Slater determinant states:

(i) *Domain wall* configuration with the left half of the chain filled and the right half empty $|\cdots 111000\cdots\rangle$. In order to quantify localization in this case we measure the total number of particles in the right half of the system (which is empty in



A. Smith, *Disorder-Free Localization*, Springer Theses,
https://doi.org/10.1007/978-3-030-20851-6_3

the initial state),

$$N_{\text{half}}(t) = \sum_{j \in \text{right half}} \langle \Psi | \hat{n}_j(t) | \Psi \rangle, \tag{3.1}$$

which tells us how many particles make it across the domain wall. This observable, as well as the long-time fermion density distribution, reveals the extent to which the fermions are localized. A similar measurement was used to identify the MBL transition in a 2D cold atom experiment in Ref. [3], and theoretically as a dynamical measure of localization in Ref. [4];

(ii) *Charge density wave* described by fermions in a Fock state with occupation numbers $|\cdots 1010 \cdots\rangle$. We will probe the memory of this initial state via the average nearest-neighbour density imbalance

$$\Delta\rho(t) = \frac{1}{\widetilde{N}} \sum_j |\langle \Psi | \hat{n}_j(t) - \hat{n}_{j+1}(t) | \Psi \rangle|, \tag{3.2}$$

where $\widetilde{N} = N - 1, N$, for open and periodic boundary conditions, respectively. This measure was used, e.g., to identify the MBL transition in cold atom experiments, see Refs. [5, 6];

(ii) *Fermi-sea* at half-filling, i.e., the ground state of the Hamiltonian

$$\hat{H}_{\text{FS}} = -J \sum_j \left( \hat{f}_j^\dagger \hat{f}_{j+1} + \text{H.c.} \right), \tag{3.3}$$

which for our polarized state is equivalent to our Hamiltonian (2.4) with $h = 0$.

Let us first consider the domain wall configuration, the results for which are shown in Fig. 3.1. We see that there is an initial ballistic spreading of fermions into

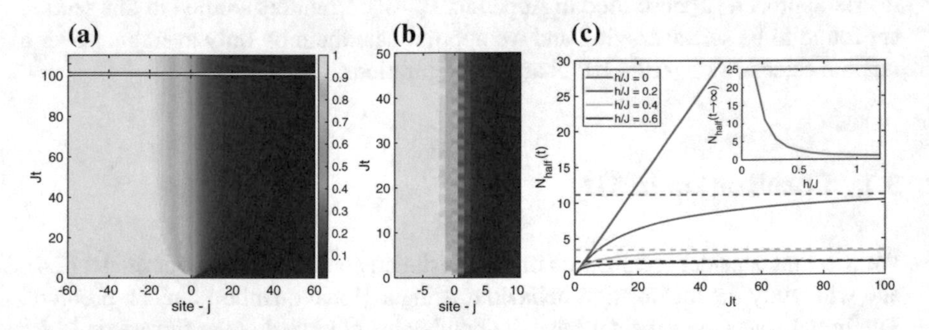


**Fig. 3.1** Time evolution of the fermion subsystem after a global quench from a domain wall initial state. **a** $\hat{n}_j(t)$ for $h/J = 0.3$ as a function of site $j$ and time $t$. Yellow indicates filled sites and blue empty. The upper panel shows the long time limit $Jt = 10^9$. **b** $\hat{n}_j(t)$ for $h/J = 2$. **c** $N_{\text{half}}(t)$ for several values of *h*/*J*. Dashed lines indicate the long time asymptotic value. (inset) The asymptotic value $N_{\text{half}}(t \to \infty)$ as a function of *h*/*J*. All results are computed for systems with $N = 200$ sites. The figure **c** is reproduced from the journal Physical Review B [7]

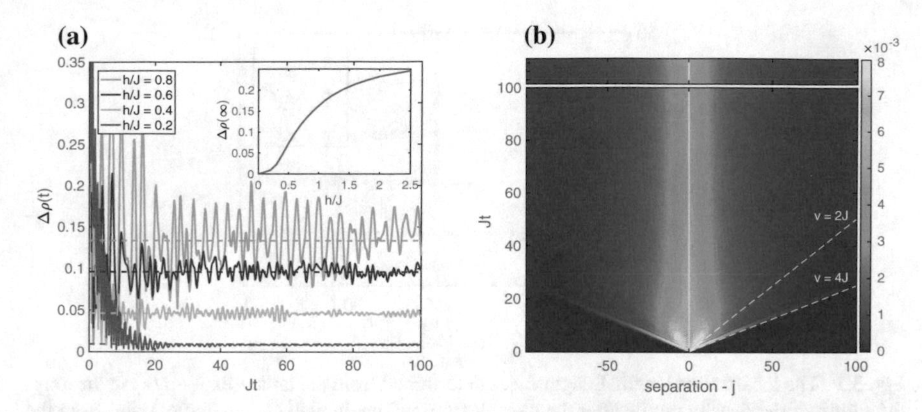


**Fig. 3.2** Time evolution of the fermion subsystem after a global quench from a domain wall initial state. **a** The neighbouring site density imbalance $\Delta\rho(t)$ for various values of *h*/*J*. The dashed lines indicate the long time value $\Delta\rho(t \to \infty)$. (inset) The long time value $\Delta\rho(t \to \infty)$ as a function of *h*/*J*. **b** The absolute value of the connected correlator $\langle\Psi|\,\hat{n}_l(t)\hat{n}_{l+j}(t)\,|\Psi\rangle_c$. The dashed lines indicate the Lieb-Robinson velocity $v_{LR} = 4J$ and a second signal at $v = 2J$. The upper panel in **b** shows the long time limit $Jt = 10^9$. The figure is reproduced from the journal Physical Review Letters [8]

the empty half of the system but this eventually halts. The number of particles that make it across the domain wall, $N_{\text{half}}$, is shown in Fig. 3.1c and grows only to a finite value, indicated by dashed lines and shown in inset as a function of *h*/*J*. The case $h/J = 0$ is included as a comparison and shows the ballistic behaviour $N_{\text{half}}(t) \propto t$. As we increase the effective disorder strength *h*/*J* the number of particles that make it across the domain wall decreases and the extent of the spreading is reduced, as shown in Fig. 3.1b. Note that as *h*/*J* is increased we also see oscillations in the fermion density back and forth across the domain wall. This is a consequence of the binary nature of the disorder, as will be discussed in detail below.

Next, let us consider the charge density wave initial state. In Fig. 3.2a we see that the average density imbalance $\Delta\rho$, defined in Eq. (3.2), starts at 1 and decays to a finite asymptotic value. This should be contrasted with the case $h/J = 0$, which was discussed in Sect. 1.2.1 of the introduction, where we found that $\Delta\rho(t \to \infty) = 0$. Here we see that the asymptotic value $\Delta\rho(t \to \infty)$ grows monotonically with *h*/*J* and is always non-zero. As the effective disorder strength is increased we also observe larger amplitude and longer lived fluctuations about this asymptotic value, similar to the oscillations observed for the quench from a domain wall.

The localization behaviour can also be diagnosed by the spreading of density correlations, shown in Fig. 3.2b. Here we consider the connected density correlator $\langle\Psi|\,\hat{n}_j(t)\hat{n}_k(t)\,|\Psi\rangle_c$, which we introduced in Sect. 1.2.1. For $h = 0$ we find linear spreading of correlations with velocity $v_{\text{LR}} = 4J$, see Fig. 1.1. For $h/J > 0$, we also observe this linear light-cone behaviour for short times but the spreading eventually halts, as shown in Fig. 3.2b for $h/J = 0.2$, which is evidence that the fermions are

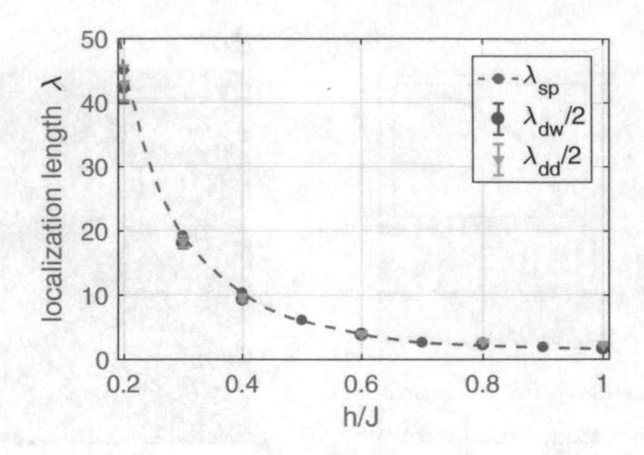


**Fig. 3.3** The localization length. Lengthscales determined from the tails $\sim\exp(-j/\lambda)$ in the long-time limit of the density profile after the quench from a domain wall ($\lambda_{dw}$—circles), and from the spatial distribution of correlations ($\lambda_{dd}$—triangles). These are compared with the single-particle localization length ($\lambda_{sp}$) [2]. The error bars are given by 2.5 standard deviations of the numerical exponential fit. For $h/J = 0.2, 0.3$ we used $N = 400$, with $N = 200$ for all others. The figure is reproduced from the journal Physical Review Letters [8]

localized. Note that at short times we are also able to identify a propagating signal at $v_{\mathrm{LR}}/2$.

For both the density profile after quenching from a domain wall and for the density-correlations starting from a charge density wave, we find that the spreading halts and we have a stationary form at long times. In both cases this long-time density or correlation profile has exponential tails (see for example, Fig. 3.4b), which are determined by the single particle localization length, with a proportionality constant of approximately 2. This is shown in Fig. 3.3 where the single particle localization length is compared with that extracted from fits to the exponential tails. The single particle localization length $\lambda_{\mathrm{sp}}$ is calculated using the spectral formula [2, 9]

$$\frac{1}{\lambda_{\mathrm{sp}}} = \min_{E} \int_{-\infty}^{\infty} g(x) \log|E - x| \, \mathrm{d}x, \tag{3.4}$$

where $g(x)$ is the DOS calculated via the kernel polynomial method, see Appendix D.

Let us now consider the quench from the Fermi-sea initial state. Unlike the domain wall and charge density wave, this state is homogeneous and we cannot probe localization using measures of density imbalance. Instead we look at the density correlations, shown in Fig. 3.4. Here we also see the initial ballistic spreading with the same Lieb-Robinson velocity $v_{\mathrm{LR}} = 4J$, which again halts and approaches a stationary form. Here we also find that the long-time exponential tails are determined by the single particle localization length, as shown in Fig. 3.4b. This figure also shows long-range density correlations in the initial state that decay with a power-law in the separation, which were not present for the CDW. These initial correlations display spatial oscillations which are determined by the Fermi-wavelength at half-filling.

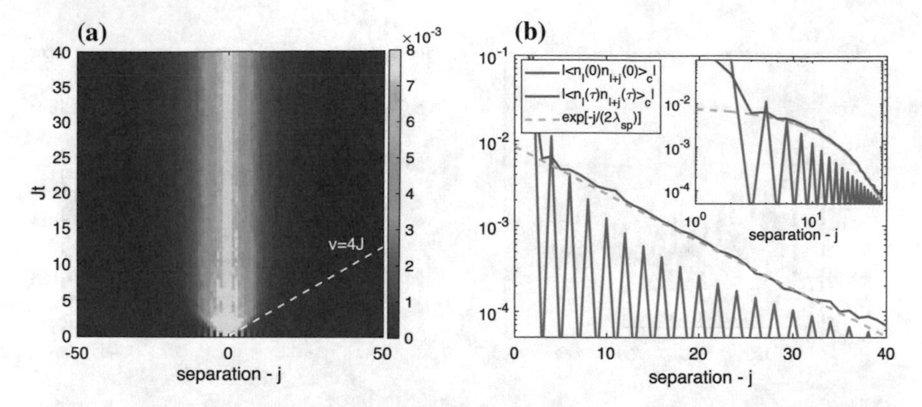


**Fig. 3.4** Connected density-density correlator after a quench from a translationally invariant Fermi-sea initial state with $h/J = 0.6$, $N = 200$ sites and periodic boundary conditions. **a** Absolute value of the connected density-density correlator $\langle 0|\hat{n}_l(t)\hat{n}_{l+j}(t)|0\rangle_c$ as a function of separation $j$ and time $t$. **b** Semi-log plot of the spatial correlator for $t = 0$ and $t = \tau = 100/J$. The dashed line is the exponential fit $\exp\{-j/(2\lambda_{sp})\}$, where $\lambda_{sp}$ is the single particle localization length. (inset) Same data on a log-log plot. The figure is reproduced from the journal Physical Review Letters [8]

### 3.1.1 Spin Subsystem

Let us now turn to the spin subsystem. We first consider the expectation value of the $z$ component of the bond spins, which are initially equal to one for all spins, see Fig. 3.5a. This magnetization decays to zero for all values of $h \neq 0$. Furthermore, for the explored range of parameters $h/J$, we find that this decay is asymptotically a power-law. As was shown in Sect. 2.2, the average magnetization can be written as

$$\langle\Psi|\hat{\sigma}^z_{jk}(t)|\Psi\rangle = \frac{1}{2^{N-1}}\sum_{\{q_i\}=\pm 1}{}' \langle\psi|e^{i\hat{H}(q)t}e^{-i\hat{H}_{jk}(q)t}|\psi\rangle, \tag{3.5}$$

which is a disorder-averaged Loschmidt echo with a local on-site potential quench between forward and backward evolution. The power-law decay is consistent with the analysis in Ref. [10], concerning the asymptotic behaviour of the Loschmidt echo in a disordered system. We find that while the short to intermediate time behaviour is non-universal as we vary $h/J$, we observe asymptotic power-law decay at long-times in all cases, even away from the perturbative regime considered in Ref. [10]. Note that in Fig. 3.5a we have shown the exact results for $N = 20$ compared with the disorder averaged results for $N = 200$ which show a remarkable qualitative agreement. This suggests that the spin dynamics is dominated by regions of finite size, presumably of the order of the fermion localization length.

We also consider the connected equal-time spin correlator $\langle\Psi|\hat{\sigma}^z_l(t)\hat{\sigma}^z_{l+j}(t)|\Psi\rangle_c$, shown in Fig. 3.5b, which exhibits an initial linear ballistic light-cone spreading. After this initial spreading we find that all spatial correlations decay to zero. The observed

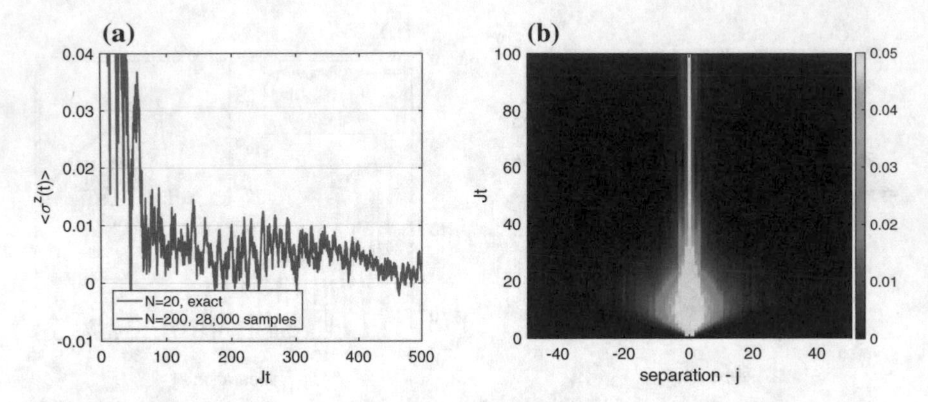


**Fig. 3.5** Time evolution of the spin subsystem. **a** The spin average $\langle\hat{\sigma}^z(t)\rangle$ of the bond-spin at $h/J = 1$ after a quench from an initial half-filled Fermi-sea state, comparing exact result for $N = 20$ with disorder averaged result for $N = 200$. **b** absolute value of the connected spin-correlator $\langle\hat{\sigma}^z_l(t)\hat{\sigma}^z_{l+j}(t)\rangle_c$ for $h/J = 0.3$, $N = 100$. The figure is reproduced from the journal Physical Review Letters [8]

decay of magnetisation and decay of correlations is suggestive of the equilibration of the spin subsystem.

### *3.1.2 Spectral Properties*

It is instructive to look at the density of states for our model which explains some of the features seen in the local expectation values and correlators above. In Fig. 3.6 we show the density of states (DOS) for a range of *h*/*J* in a fixed charge sector, shifted so that the band center is $E = 0$. The most striking qualitative change as $h$ is increased is how 'spiky' the DOS becomes at the band edges, and across the entire bandwidth for $h \gg J$. This spiky behaviour leads to long lived resonances which show up as large fluctuations in the observables, as we saw in Figs. 3.1b and 3.2a.

When we take $h/J > 1$ we also observe that the DOS splits up into two bands centered on $\pm 2h$, each with bandwidth $4J$. Note that in the Falicov-Kimball model this corresponds to the Mott phase [11]. This can be interpreted as the separation of eigenstates into those living at the top and bottom of the binary potential. This property is due to the binary nature of the effective disorder and has some striking consequences for the localization behaviour in higher dimensions which will be discussed in the subsequent sections.

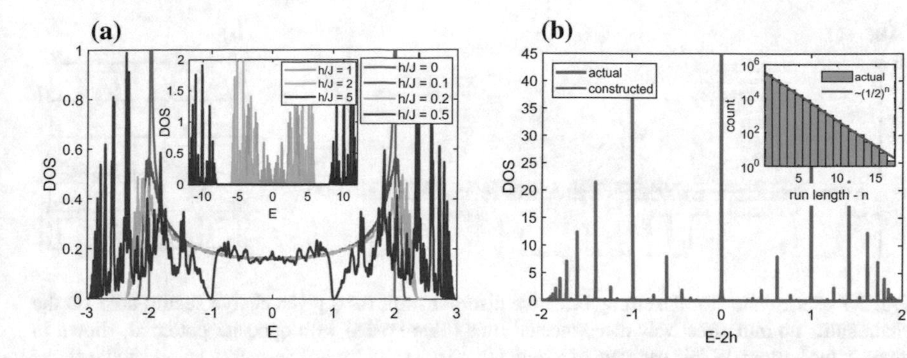


**Fig. 3.6** Density of states for the 1D chain (2.22) with fixed random charge configuration. **a** DOS for different values of *h*/*J*. (inset) The DOS for values of $h > J$ (where there is a gap in the DOS). **b** DOS for a large value of $h/J = 500$. The energy is offset by $2h$ and we focus on one of the two sub-bands that form for large $h$. The DOS is computed using the kernel polynomial method (see Appendix D), shown in blue. We compare this with the DOS constructed using Eq. (3.7), shown in red. (inset) A comparison of the observed distribution of run lengths $n$ with the corresponding distribution in the thermodynamic limit $\sim(1/2)^n$. The figure is reproduced from the journal Physical Review B [7]

## 3.2 Strong Effective Disorder Limit

As can be seen in the DOS in Fig. 3.6b, the spectrum changes significantly for large effective disorder *h*/*J*. This behaviour is unique to systems with binary disorder and leads to several observable effects. See for example Refs. [12, 13] for further discussion of the differences between binary, higher-order discrete, and continuously sampled disorder. In the following sections we will also demonstrate that this strong disorder limit can lead to an effective quantum percolation problem, which can result in the delocalization of the fermions.

In this section we will pay particular attention to the large *h*/*J* limit of our model using perturbative arguments. For a given charge configuration in this limit the system effectively splits into a collection of finite-length runs, which are defined to be those connected sections where the charges have the same sign, see Fig. 3.7. In the thermodynamic limit $N \to \infty$, the distribution of the length $l$ of the runs is given by the geometric distribution $\sim(1/2)^l$, as demonstrated in the inset of Fig. 3.6b. In this limit it becomes natural to separate the Hamiltonian into $\hat{H}_h$ and $\hat{H}_J$ given by

$$\begin{aligned} \hat{H}_h &= 2h \sum_i q_i (\hat{c}_i^\dagger \hat{c}_i - 1/2) - J \sum_{\langle ij \rangle : q_i = q_j} (\hat{c}_i^\dagger \hat{c}_j + \text{H.c}), \\ \hat{H}_J &= -J \sum_{\langle ij \rangle : q_i = -q_j} (\hat{c}_i^\dagger \hat{c}_j + \text{H.c}). \end{aligned} \tag{3.6}$$

Here the sums over nearest-neighbours are only between sites on which the charges are the same sign in $\hat{H}_h$, and only between those of differing sign in $\hat{H}_J$. Note that we omit an overall energy shift $h \sum_i q_i$ which does not affect the results since

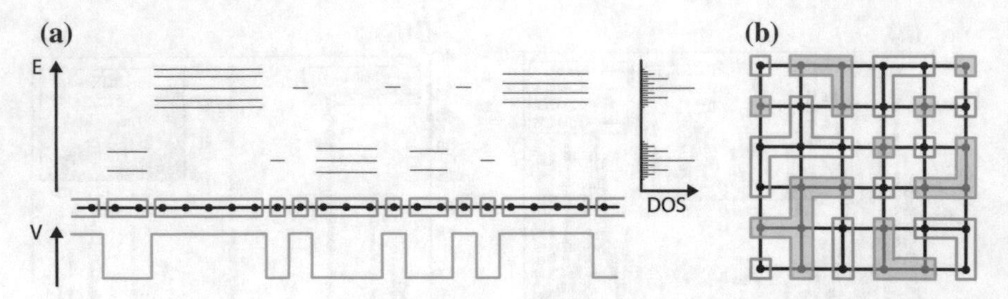


**Fig. 3.7** Schematics for the strong effective disorder limit for a given charge sector. **a** In 1D the chain splits up into effectively disconnected runs (blue boxes) with opposite potential, shown in green at the bottom. Within each run of length $l$ there are $l$ single particle energy levels shown above in red. The combination gives the DOS (right) shown in Fig. 3.6. **b** Example of disconnected regions in 2D with clear regions indicating low potential and shaded regions indicating high potential

there are no matrix elements between different charge sectors. The Hamiltonian $\hat{H}_h$ describes disconnected, uniform tight-binding chains, and $\hat{H}_J$ corresponds to a hopping between these chains, see Fig. 3.7a. Note that the Hamiltonian takes the same form on any lattice in any dimensions—see Fig. 3.7b for a schematic picture of the disconnected region in 2D.

In our perturbative arguments we take as our starting point the Hamiltonian $\hat{H}_h$, that is, an ensemble of disconnected runs of random length. In each isolated run of length $l$, we then have $l$ single particle eigenfunctions and energy levels $E$. The DOS of the Hamiltonian $\hat{H}_h$ can be constructed using an ensemble of the energy levels for disconnected chains weighted by their probability distribution using the following equation

$$g(\omega) \propto -\frac{1}{\pi} \operatorname{Im} \lim_{\delta \to 0} \sum_{l=1}^{\infty} \sum_{E_l} \frac{(1/2)^l}{\omega - E_l + i\delta}, \tag{3.7}$$

where $E_l$ denote the single-particle eigenvalues of the tight-binding Hamiltonian for a uniform chain of length $l$, see Fig. 3.7a. In order to obtain the DOS numerically we introduce a cutoff on the sum over $l$ and choose a finite broadening $\delta = 0.0015$. The spectrum splits into two sub-bands centred at $\pm 2h$ corresponding to the two types of run with $q = \pm 1$, as seen in inset of Fig. 3.6a. In Fig. 3.6b we compare the exact DOS centred around one of these sub-bands at $E = 2h$ for a large but finite system with large $h \gg J$, and the DOS constructed from Eq. (3.7), which shows good agreement.

The fact that the DOS splits up into a discrete set of levels, of which only a few carry the majority of the spectral weight, explains the observed fluctuations in our localization diagnostics. These features can be attributed to resonant processes between these few discrete levels. Figure 3.6a shows that a similar structure persists, to some extent, below $h/J = 1$ in 1D. The effect of $\hat{H}_J$ on the DOS appears at second order in perturbation theory.[1] This gives rise to the broadening of the spectrum,

[1]The first order is zero since the overlap of eigenstates between separate runs is zero.

and provides a time-scale $\sim(h/J)^2$ which sets the lifetime of the fluctuations in the observables. This time scale will be particularly evident when we look at the entanglement properties in Chap. 4.

## 3.3 Localization in 2D

In two dimensions we will study the Hamiltonian (2.1) on a square lattice, see Fig. 2.1. As in 1D we consider initial states with spins polarized along the $z$-axis. Here we prepare the fermions in one of the three following initial configurations, shown schematically in Fig. 3.8:

(i) *Charge density wave* with alternating occupation along one of the directions of the lattice and uniform occupation along the other (stripes), as considered experimentally in Ref. [6];
(ii) *Checkerboard pattern* with alternating occupation along both directions of the lattice;
(iii) *Domain wall* configuration with one half of the system filled, and the other empty, such as studied in cold-atom experiments, see Ref. [3].

For all diagnostics, we measure correlators only along a 1D cut through the system, e.g., perpendicular to the domain wall, as shown in Fig. 3.8. We can therefore use density imbalance measures $\Delta\rho$ and $N_{\text{half}}$ that we considered in 1D.

As in 1D we find that the average density imbalance $\Delta\rho(t)$ saturates at a non-zero value at long times, see Fig. 3.9a. However, compared to 1D, the localization length is larger in 2D (for the same $h/J$) leading to smaller long-time values for $\Delta\rho$ and larger values for $N_{\text{half}}$. Furthermore, for the values of $h/J$ shown in Fig. 3.9, which are much larger than those presented for the 1D case, the amplitude of the fluctuations is much smaller. In other words, the extra dimensionality produces a damping effect on these fluctuations, which is reflected in the much smoother single-particle DOS, even for $h/J > 1$, shown in Fig. 3.10a. We also find that for the checkerboard initial state, the remaining density imbalance $\Delta\rho(t \to \infty)$ is greater than for the charge

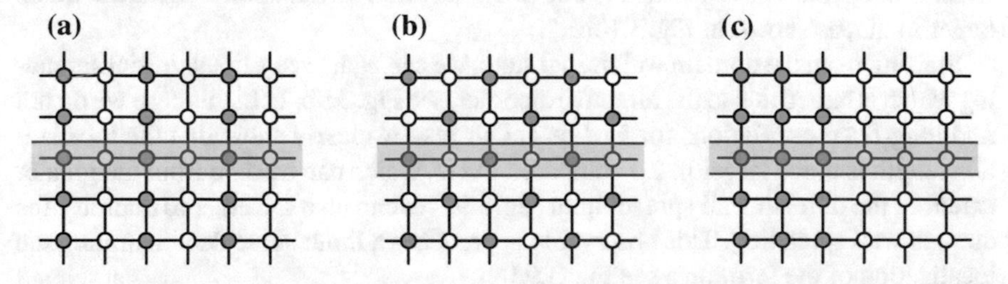


**Fig. 3.8** Initial states for the global quench in 2D: **a** charge density wave, **b** checkerboard, **c** domain wall. The filled sites are indicated in blue and empty in white. Measurements are made along the 1D strip indicated by the red box

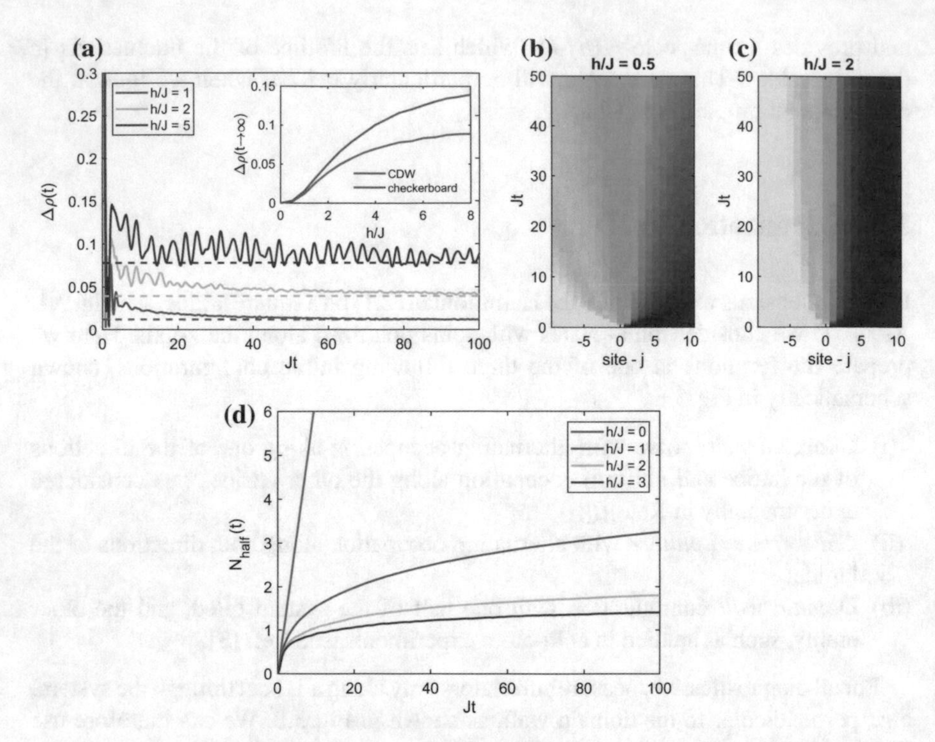


**Fig. 3.9** Time evolution of the fermion subsystem in 2D. **a** Density imbalance $\Delta\rho$ measured along a slice through the centre of the system with the initial state described by a charge density wave, see text. Inset shows the long-time limit for the charge density wave, and checkerboard initial states. **b–c** Spreading of the domain wall for $h/J = 0.5$ and $h/J = 2$, respectively, measured along the slice through the centre of the system. **d** Number of particles $N_{\text{half}}$ along the centre in the initially empty half of the system. In **a–c** we use a square lattice with $N = 32 \times 32$ sites and in **d** $N = 30 \times 50$. Results are computed using the determinant method of Appendix B. The figure is reproduced from the journal Physical Review B [7]

density wave, as shown in the inset. Comparison of the corresponding 1D and 2D results shows that the remaining imbalance is approximately an order of magnitude smaller in 2D than in 1D, which is due to the fact that localization lengths are much larger in 2D, as shown in Fig. 3.10b.

Starting from the domain wall initial state, we can again see a linear initial spreading which is halted due to the effective disorder, see Fig. 3.9b–c. In this case we do not find long-lived oscillations for $h/J > 1$. Our results clearly show that the localization length is much larger in 2D compared to 1D, which can be seen from the greater extent of the domain wall spreading in Fig. 3.9. We can also use $N_{\text{half}}$ to quantify the domain wall spreading. This observable approaches a finite value demonstrating the localization of the fermions, see Fig. 3.9d.

If we compare the DOS in 2D with that of 1D we notice some important similarities and differences. First, we see a gap opening in both cases for large values of $h$. Whereas in the 1D case this gap appears at $h = J$, the spectrum is still gapless for

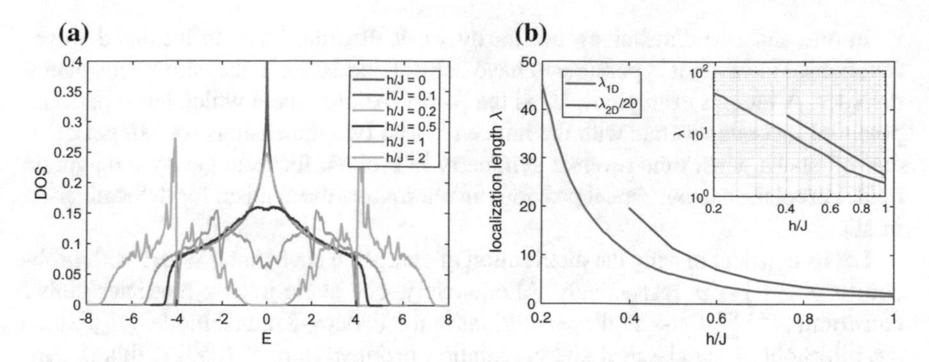


**Fig. 3.10** **a** Single-particle density of states for a 2D square lattice for different *h*/*J*. The DOS is computed using the kernel polynomial method, see Appendix D. **b** Localization length in 1D and 2D. In 1D, localization length is computed using the spectral formula (3.4) as explained in the main text. In 2D, the localization length is scaled by a factor of 20 and is computed by the transfer matrix method on a strip of width 100 sites and length 250,000 sites. The figure is reproduced from the journal Physical Review B [7]

$h \sim J$ in 2D. There is also an increase in the bandwidth in 2D, both of the total DOS and of the individual sub-bands that develop in the large $h$ limit, due to the extra dimension. More importantly we find that the sub-bands remain much smoother than in 1D for a much wider range of effective disorder strength.

In Fig. 3.10 we show the dependence of the maximum localization length on $h/J$ in 1D and 2D. The localization length $\lambda$ is the characteristic length scale of the exponential tails of the single-particle wavefunctions defined via $e^{-j/\lambda}$. In the 1D case the results are obtained using the spectral formula [2, 9] in Eq. (3.4) and in the 2D case we used the transfer matrix method [2], see Appendix E. The 2D results are rescaled by a factor of 20 which demonstrates an order of magnitude difference in localization lengths in 1D and 2D. However, the localization length as a function of disorder strength shows similar power-law in 1D as in 2D, see Fig. 3.10b inset.

## 3.4 Quantum Percolation

Due to the binary nature of the effective disorder potential in our model, we can make a connection to quantum percolation. In the large $h$ limit of strong disorder, discussed in Sect. 3.2, the lattice is decomposed into parts with sites sitting at the top or bottom of the binary potential. To the lowest-order, only hopping between sites with the same sign of the potential is allowed, see Fig. 3.7b. It is then natural to decompose the Hamiltonian as in Eq. (3.6). In $\hat{H}_h$, there will only be hopping terms between neighbouring sites with the same values of $q$, which defines a quantum site percolation problem [12], see Fig. 3.11.

In one and two dimensions, arbitrarily weak disorder leads to localized wavefunctions. However, it is possible to have delocalized states in the case of correlated disorder. A famous example in 1D is the Aubry-Andre model which has a periodic potential incommensurate with the lattice [14]. In two dimensions we can get delocalized states when time-reversal symmetry is broken, for example by a magnetic field. Percolation in our model provides an alternative mechanism for delocalization in 2D.

Let us consider biasing the distribution of charges $q$ such that $q = \pm 1$ with probability $p$ and $1 - p$, respectively. Alternatively, one could impose a stricter global constraint $N^{-1} \sum_i (\hat{q}_i + 1)|\Psi\rangle = p|\Psi\rangle$, as in the Falicov-Kimball model [15]. Since the threshold in the classical site percolation problem is $p_c \approx 0.5927$, this is consistent with having localized wavefunctions for all $h$ for our polarized spin state which corresponds to $p = 1/2$. However, if we set $p > p_c$, or $1 - p > p_c$ then we might expect percolation in our model for large $h$, see Fig. 3.11. Note that if we have delocalized wavefunctions, say at the top of the potential, then we necessarily have localized wavefunctions at the bottom of the potential, and vice versa.

In order to understand the effect of percolation we study the time evolution from a domain wall initial state with changing system size and bias $p$, as shown in Fig. 3.12a. We plot $N_{\text{half}}(t \to \infty)/L$, where $L$ is the linear dimension of a square lattice with

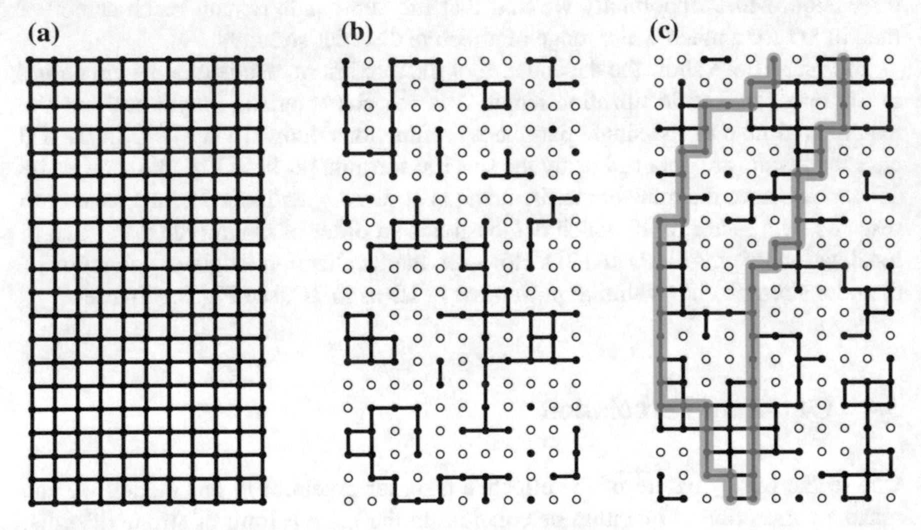


**Fig. 3.11** Schematic picture of the site percolation problem. **a** A fully connected lattice on which fermions can hop in the limit $h/J = 0$. **b–c** When $h/J \gg 1$ the fermions are constrained to hopping only between sites with the same effective potential (filled sites and bonds) and the other sites become inaccessible (open circles) leading to a quantum site percolation problem. **b** Connected sites for the bias $p = 0.5$ showing the absence of paths (in the thermodynamic limit) that connect opposite sides of the lattice. **c** Connected sites for the bias $p = 0.7$, which is above the classical percolation threshold $p_c \approx 0.5927$, and which has connected paths across the system, examples of which are shown in red. The figure is reproduced from the journal Physical Review B [7]

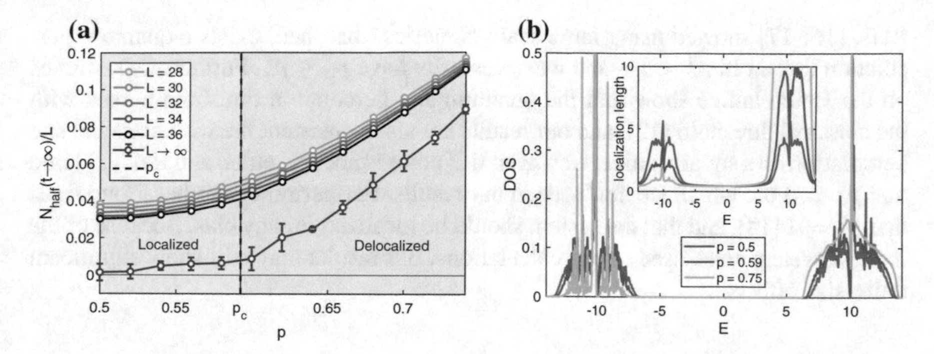


**Fig. 3.12** **a** Total number of particles making it across a domain wall $N_{\text{half}}$ for various values of the charge distribution bias, $p$. The figure shows the rescaled value $N_{\text{half}}/L$, and the extrapolated limit $L \to \infty$, where $L$ is the linear dimension of the system. Error bars correspond to 2 standard deviations in the linear fit from finite size scaling. Note that $p_c \approx 0.5927$ corresponds to the classical site percolation threshold for a square lattice. **b** Density of states for different values of the bias $p$ and $h/J = 20$. Inset shows the corresponding energy-resolved localization length. The results shown in **a** were obtained using the determinant method (Appendix B). We used the KPM (Appendix D) for the DOS in **b** and the transfer matrix approach for the localization length (Appendix E) for a system of $150 \times 250,000$ sites. The figure is reproduced from the journal Physical Review B [7]

$N = L \times L$ sites.[2] If particles are localized then we would expect $N_{\text{half}}(t \to \infty)/L$ to tend to zero as $L$ is increased—i.e., $N_{\text{half}}$ is finite and independent of $L$. Whereas, if the particles are delocalized, we should find that a finite proportion of the particles makes it across the domain wall and $N_{\text{half}}(t \to \infty)/L \to$ constant. In Fig. 3.12a we show the extrapolation from finite size scaling which agrees with this expected behaviour in the two limits. Furthermore, the change in behaviour is observed to happen around the classical percolation threshold $p = p_c$.

This percolation behaviour is also reflected in the DOS and the energy-resolved localization length as a function of $p$, as shown in Fig. 3.12b. As $p$ is increased past the critical point we see a clear asymmetry with respect to energy in these results. In the DOS, one of the sub-bands becomes similar to that of the large $h$ limit of the $p = 1/2$ problem, characterized by a discrete set of levels with large spectral weight. The other sub-band become much smoother and similar to the DOS for a clean 2D system, see Fig. 3.10. Furthermore, we respectively see a decrease and increase in the localization length in the these two sub-bands, see inset, which is consistent with percolation of the fermions at positive potential and localization of those at negative potential.

Because we are studying quantum dynamics, it is not clear that there should be a direct correspondence with the classical site percolation problem. In particular, given a path through the system, as in Fig. 3.11c, we would generally expect quantum fluctuations to lead to backscattering, which may hinder conductance. However,

[2]Note that in our definition of $N_{\text{half}}$ we only sum over the initially empty sites in the 1D strip shown in Fig. 3.8c.

Refs. [16, 17] showed using large scale numerics, that there exists a quantum percolation threshold $p_c^q < 1$—and we necessarily have $p_c \leq p_c^q$. Furthermore, studies on the Bethe lattice show that the quantum site percolation threshold agrees with the classical threshold [12], and our results are also consistent the with classical site percolation on a square lattice. However, this point is not yet settled and Ref. [17] find that $p_c^q > 0.65$, which we don't see in our results. Furthermore, another group finds that $p_c^q = 1$ [18], and that the system should be localized for any bias. Because of the modest system sizes used in our calculations, our results may still show significant finite-size effects.

### *3.4.1 Localization in 3D*

For a 3D cubic lattice Ref. [17] find that $p_c^q < 0.5$. Therefore the 3D version of the model offers a particularly interesting setting for studying localization. In 3D there is a critical disorder strength needed to achieve localization. However, based on the results of Ref. [17], in the large $h/J$ limit we would expect delocalized states to exist for all values of the bias probability $p$. This then raises the possibility of delocalized states for both low and high $h/J$ but localized states for intermediate values. A careful investigation of the 3D model goes beyond this thesis.

## References


1. Weiße A, Wellein G, Alvermann A, Fehske H (2006) The kernel polynomial method. Rev Mod Phys 78:275–306. https://doi.org/10.1103/RevModPhys.78.275
2. Kramer B, MacKinnon A (1993) Localization: theory and experiment. Rep Prog Phys 56:1469–1564. https://doi.org/10.1088/0034-4885/56/12/001
3. Choi J-y, Hild S, Zeiher J, Schauss P, Rubio-Abadal A, Yefsah T, Khemani V, Huse DA, Bloch I, Gross C (2016) Exploring the many-body localization transition in two dimensions. Science (80-.) 352:1547–1552. https://doi.org/10.1126/science.aaf8834
4. Hauschild J, Heidrich-Meisner F, Pollmann F (2016) Domain-wall melting as a probe of many-body localization. Phys Rev B 94:161109. https://doi.org/10.1103/PhysRevB.94.161109
5. Schreiber M, Hodgman SS, Bordia P, Luschen HP, Fischer MH, Vosk R, Altman E, Schneider U, Bloch I (2015) Observation of many-body localization of interacting fermions in a quasirandom optical lattice. Science (80-.) 349:842–845. https://doi.org/10.1126/science.aaa7432
6. Bordia P, Lüschen H, Scherg S, Gopalakrishnan S, Knap M, Schneider U, Bloch I (2017) Probing slow relaxation and many-body localization in two-dimensional quasiperiodic systems. Phys Rev X 7:041047. https://doi.org/10.1103/PhysRevX.7.041047
7. Smith A, Knolle J, Moessner R, Kovrizhin DL (2018) Dynamical localization in $Z_2$ lattice gauge theories. Phys Rev B 97:245137. https://doi.org/10.1103/PhysRevB.97.245137
8. Smith A, Knolle J, Kovrizhin DL, Moessner R (2017) Disorder-free localization. Phys Rev Lett 118:266601. https://doi.org/10.1103/PhysRevLett.118.266601
9. Thouless DJ (1972) A relation between the density of states and range of localization for one dimensional random systems. J Phys C Solid State Phys 5:77–81. https://doi.org/10.1088/0022-3719/5/1/010

10. Vardhan S, De Tomasi G, Heyl M, Heller EJ, Pollmann F (2017) Characterizing time irreversibility in disordered fermionic systems by the effect of local perturbations. Phys Rev Lett 119:016802. https://doi.org/10.1103/PhysRevLett.119.016802
11. Antipov AE, Javanmard Y, Ribeiro P, Kirchner S (2016) Interaction-tuned anderson versus mott localization. Phys Rev Lett 117:146601. https://doi.org/10.1103/PhysRevLett.117.146601
12. Alvermann A, Fehske H (2005) Local distribution approach to disordered binary alloys. Eur Phys J B 48:295–303. https://doi.org/10.1140/epjb/e2005-00408-8
13. Janarek J, Delande D, Zakrzewski J (2018) Discrete disorder models for many-body localization. Phys Rev B 97:155133. https://doi.org/10.1103/PhysRevB.97.155133
14. Aubry S, André G (1980) Analyticity breaking and anderson localization in incommensurate lattices. Ann Isreal Phys Soc 3:18
15. Falicov LM, Kimball JC (1969) Simple model for semiconductor-metal transitions: SmB6 and transition-metal oxides. Phys Rev Lett 22:997–999. https://doi.org/10.1103/PhysRevLett.22.997
16. Schubert G, Fehske H (2008) Dynamical aspects of two-dimensional quantum percolation. Phys Rev B 77:245130. https://doi.org/10.1103/PhysRevB.77.245130
17. Schubert G, Fehske H (2009) Quantum percolation in disordered structures. arXiv:0905.2537
18. Chandrashekar CM, Busch T (2015) Quantum percolation and transition point of a directed discrete-time quantum walk. Sci Rep 4:6583. https://doi.org/10.1038/srep06583

# Chapter 4
# Entanglement Properties

In this chapter we consider the entanglement properties of our model. In particular we discuss how our model fits into the recently proposed notion of a quantum disentangled liquid (QDL) [1]. The QDL was motivated by the heavy-light mixture models [2, 3] and constitutes a novel phase of matter for multi-component liquids. The loose idea is that some components 'thermalize' while the others are localized. The QDL is defined in terms of projective measures of entanglement which reveal contrasting behaviours for the different components of the model. We apply this diagnostic to the long time state of our model after a quantum quench. We find that these measures are not able to identify the localization behaviour of the $f$ fermions in our model, but that they can for the $c$ fermions in the dual language. This demonstrates the subtlety in defining a QDL.

To shed some light on these projective measures of entanglement we deconstruct their definition. While our investigation does reveal a distinction between species, it does not provide a conclusive and universal signature for the QDL. We also investigate the quantum mutual information as an alternative entanglement measure, which successfully distinguishes the localized behaviour of the fermions from that of the spin subsystems.

## 4.1 Quantum Disentangled Liquids

The study of heavy-light mixtures [2, 3] brought forward the notion of a quantum disentangled liquid (QDL) [1]. This is a proposed phase of matter which is defined in terms of projective measures of entanglement. In a thermalizing system, dynamics after a quench induces growth of the von Neumann entanglement entropy which saturates with volume law scaling, whereas in a localized system we have area law entanglement of all eigenstates. The broad idea of a QDL is that it has multiple


A. Smith, *Disorder-Free Localization*, Springer Theses,
https://doi.org/10.1007/978-3-030-20851-6_4

subsystems, some of which 'thermalize' in this entanglement sense, and importantly, others that do not.

The QDL was defined in the context of multicomponent liquids [1] but we restrict ourselves to consider only two-components, which is the situation in our model, and for the heavy-light particle mixtures. The main tools in defining a QDL are the projective measures of entanglement, which we call the projective bipartitie entanglement entropy (PBEE), defined in the following section. The QDL phase has eigenstates with volume-law scaling of the von Neumann entanglement entropy and the PBEE for one of the species, $S^{\alpha}_{\text{PBEE}}$, but area-law scaling for the other species $S^{\beta}_{\text{PBEE}}$.

### 4.1.1 Projective Bipartitie Entanglement Entropy (PBEE)

To define and diagnose the QDL phase we need to introduce the PBEE. Consider a two component system, with the components labelled by $\alpha$ and $\beta$, in a pure state $|\psi\rangle$. $\hat{P}^{\gamma}_{\phi}$ is the projector onto the state $|\phi\rangle$ of species $\gamma \in \{\alpha, \beta\}$. This projector is related to a measurement of the single component. We also spatially partition our system into two subsystems $A$ and $B$. The algorithm for computing the PBEE for the component $\alpha$ (similarly for $\beta$) is then as follows:

(i) Project the state $|\psi\rangle$ onto state $|\phi\rangle$ of species $\beta$, i.e. $|\psi\rangle_{\phi} = \hat{P}^{\beta}_{\phi}|\psi\rangle / \sqrt{\mathcal{Z}^{\alpha}_{\phi}}$, where $\mathcal{Z}^{\alpha}_{\phi} = \langle\psi|\hat{P}^{\beta}_{\phi}|\psi\rangle$ normalizes the state;

(ii) Define the reduced density matrix with respect to the bipartition $\rho^{\phi}_{A} = \text{Tr}_{B}|\psi\rangle_{\phi}\langle\psi|_{\phi}$;

(iii) Compute the von Neumann entanglement entropy $S^{\alpha}_{\phi} = -\text{Tr}_{A}[\rho^{\phi}_{A} \log \rho^{\phi}_{A}]$;

(iv) The PBEE for the species $\alpha$ with respect to the bipartition is then defined as

$$S^{\alpha}_{\text{PBEE}} = \sum_{|\phi\rangle} \mathcal{Z}^{\alpha}_{\phi} S^{\alpha}_{\phi}, \tag{4.1}$$

where the sum over entropies is weighted by the probabilities of the states $|\psi\rangle_{\phi}$.

In the original paper [1] the authors illustrate the properties of QDL by constructing particular states, one in the setting of heavy-light mixtures and the other using a Born-Oppenheimer type approach to construct a wavefunction. They also hypothesise that the QDL behaviour may be observable in the Hubbard model where instead of a separation into two particle species we have a separation between charge and spin degrees of freedom. This idea was investigated further in Refs. [4–6], wherein they conclude that the low-lying "spin-band" eigenstates indeed satisfy the diagnostic for QDL. Rather than looking at eigenstates, we will be concerned with the state at long times after our global quantum quench protocol. Since we observe the localization in the fermion subsystem and relaxation of the magnetization of the spins, our model would seem to be an ideal candidate for realizing a QDL state at long times.

**Example**

Before we proceed, we will consider an explicit example of a four spin system to aid the definition of the PBEE. We divide the system equally into two partitions $A$ and $B$, each of which contains a spin of each species, and will work in the local $z$-basis for each spin, i.e. $\{|\uparrow\rangle, |\downarrow\rangle\}$. Let us consider the pure state

$$|\psi\rangle = \frac{1}{2}\left[(|\uparrow_A\downarrow_B\rangle_\alpha + |\downarrow_A\uparrow_B\rangle_\alpha)|\uparrow_A\uparrow_B\rangle_\beta + (|\uparrow_A\downarrow_B\rangle_\alpha - |\downarrow_A\uparrow_B\rangle_\alpha)|\uparrow_A\downarrow_B\rangle_\beta\right]. \tag{4.2}$$

First we will calculate $S^\alpha_{\text{PBEE}}$ for the $\alpha$ species. With respect to this basis we have four projection operators onto states of species $\beta$, which are $\hat{P}^\beta_{\uparrow\uparrow}$, $\hat{P}^\beta_{\uparrow\downarrow}$, $\hat{P}^\beta_{\downarrow\uparrow}$ and $\hat{P}^\beta_{\downarrow\downarrow}$, whose actions on the pure state $|\psi\rangle$ are given by

$$\hat{P}^\beta_{\uparrow\uparrow}|\psi\rangle = \frac{1}{2}(|\uparrow_A\downarrow_B\rangle_\alpha + |\downarrow_A\uparrow_B\rangle_\alpha), \qquad \hat{P}^\beta_{\uparrow\downarrow}|\psi\rangle = \frac{1}{2}(|\uparrow_A\downarrow_B\rangle_\alpha - |\downarrow_A\uparrow_B\rangle_\alpha), \tag{4.3}$$

and $\hat{P}^\beta_{\downarrow\uparrow}|\psi\rangle = \hat{P}^\beta_{\downarrow\downarrow}|\psi\rangle = 0$. The non-zero probabilities of measuring these states are $\mathcal{Z}^\alpha_{\uparrow\uparrow} = \frac{1}{2}$, $\mathcal{Z}^\alpha_{\uparrow\downarrow} = \frac{1}{2}$ and the density matrices for these two normalized states are then

$$\rho^{\uparrow\uparrow} = \begin{pmatrix} 0 & 0 & 0 & 0 \\ 0 & \frac{1}{2} & \frac{1}{2} & 0 \\ 0 & \frac{1}{2} & \frac{1}{2} & 0 \\ 0 & 0 & 0 & 0 \end{pmatrix}, \qquad \rho^{\uparrow\downarrow} = \begin{pmatrix} 0 & 0 & 0 & 0 \\ 0 & \frac{1}{2} & -\frac{1}{2} & 0 \\ 0 & -\frac{1}{2} & \frac{1}{2} & 0 \\ 0 & 0 & 0 & 0 \end{pmatrix}. \tag{4.4}$$

Tracing out subsystem $B$ we then get the reduced density matrices

$$\rho^{\uparrow\uparrow}_A = \begin{pmatrix} \frac{1}{2} & 0 \\ 0 & \frac{1}{2} \end{pmatrix} = \rho^{\uparrow\downarrow}_A, \tag{4.5}$$

which are the same as for the EPR state discussed in Sect. 1.2.3. We thus find that $S^\alpha_{\uparrow\uparrow} = S^\alpha_{\uparrow\downarrow} = \log 2$ and the PBEE is given by

$$S^\alpha_{\text{PBEE}} = \frac{1}{2}\log 2 + \frac{1}{2}\log 2 = \log 2, \tag{4.6}$$

which in this case is the same as the von Neumann entanglement entropy of the maximally entangled EPR pair.

Repeating this procedure for the species $\beta$ we find that the only non-zero states after projection are

$$\hat{P}^{\alpha}_{\uparrow\downarrow}|\psi\rangle = \frac{1}{2}\left(|\uparrow_A\uparrow_B\rangle_\beta + |\uparrow_A\downarrow_B\rangle_\beta\right), \qquad \hat{P}^{\alpha}_{\downarrow\uparrow}|\psi\rangle = \frac{1}{2}\left(|\uparrow_A\uparrow_B\rangle_\beta - |\uparrow_A\downarrow_B\rangle_\beta\right). \tag{4.7}$$

We therefore have that $S^{\beta}_{\uparrow\downarrow} = S^{\beta}_{\downarrow\uparrow} = 0$, since these are tensor product states, which further gives

$$S^{\beta}_{\text{PBEE}} = 0, \tag{4.8}$$

for the species $\beta$. Note that unlike this example, we cannot uniquely define an equal partition of our system because the spins live on the bonds of the lattice. Therefore, taking an equal partition of the fermion subsystem necessarily means that we have an unequal partition for the spins because of the spin on the central bond, but this does not affect the scaling behaviour.

### *4.1.2 Results*

Here we present the results of numerical simulations, which were computed using exact diagonalization methods for up to $N = 12$, with open boundary conditions. In all cases we quench from initial states with $z$-polarized spins and fermions in a charge density wave. We begin by looking at the von Neumann entropy for our model after the quench before comparing the PBEEs for the original $f$ and $\sigma$ degrees of freedom. We then also consider these projective measures in the dual language in terms of the $c$-fermions and charges $q$.

In Fig. 4.1a we show the bipartite von Neumann entanglement entropy for $h/J = 20$ after a quench from a charge density wave fermion state and $z$-polarized spins. The entropy exhibits initial linear growth followed by an area-law plateau which eventually gives way to the volume-law scaling (note the dependence on the system size). The extent of the plateau scales as $(h/J)^2$ for $h/J > 1$, as shown in the inset; it is absent for $h/J < 1$. This behaviour can be attributed to a separation of timescales. This was discussed in Sect. 3.2 where, for $h/J \gg 1$, a pair of adjacent sites with opposite values of $q_j$ correspond to a high energy barrier. Traversing such a barrier is a process parametrically suppressed in $h/J$, while motion between such barriers takes place on shorter timescales. The latter can only produce area-law scaling of the entanglement entropy, while the former can act over longer distances, resulting in equilibration of the spins and a concomitant volume-law scaling for the entanglement entropy. Note that the same two localization regimes also appear in the disorder-averaged entanglement entropy of a simple tight-binding model with binary disorder. It is directly related to the PBEE projected onto the charge sectors in our model, $S^{c}_{\text{PBEE}}$ shown in Fig. 4.2a, because our choice of spin polarized initial state leads to an equal weight superposition of all disordered charge configurations.

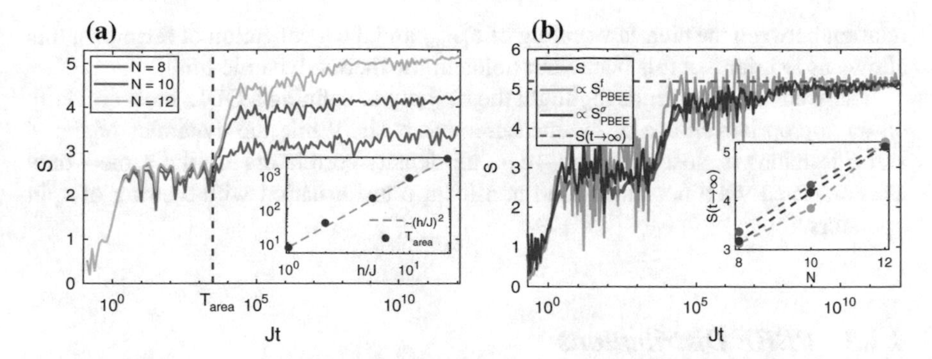


**Fig. 4.1** Time evolution of entanglement entropy after a quench from a charge density wave state. The results are obtained using exact diagonalization for $h/J = 20$. **a** The von Neumann bipartite entanglement entropy $S(t)$ for $N = 8, 10, 12$. (inset) The time $T_{\text{area}}$ for which the area-law plateau persists (dashed line of main plot) as a function of *h*/*J* compared with $(h/J)^2$. **b** Comparison between rescaled PBEEs $S^f_{\text{PBEE}}(t)$, $S^\sigma_{\text{PBEE}}(t)$, and the von Neumann entropy $S(t)$ for $N = 12$. (inset) The long time-limit $S(t \to \infty)$ (computed at $Jt \sim 10^{12}$) as a function of system size. PBEE results are scaled by factors of 2.7 and 1.95, respectively. The figure is reproduced from the journal Physical Review Letters [7]

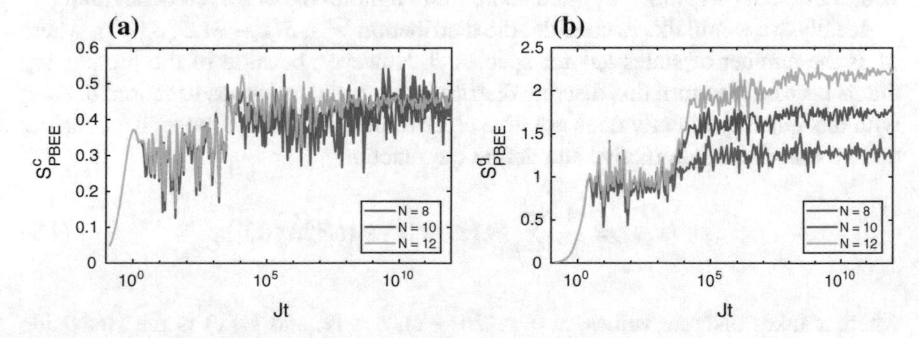


**Fig. 4.2** The time evolution of the PBEE, defined in the main text, starting from *z*-polarized spins and a charge density wave state for the fermions with $h/J = 20$, and $N = 8, 10, 12$, obtained using exact diagonalization. **a** $S^c_{\text{PBEE}}(t)$ for *c*-fermions. **b** $S^q_{\text{PBEE}}(t)$ for the conserved charges *q*. The figure is reproduced from the journal Physical Review Letters [7]

The PBEEs for the original degrees of freedom, the $f$-fermions and $\sigma$-spins, are shown in Fig. 4.1b. The data is scaled to highlight the fact that both PBEEs have the same qualitative behaviour, and match the von Neumann entanglement entropy of the composite system. In terms of the $f$ and $\sigma$ degrees of freedom, the long time limit *does not* suggest the QDL behaviour since all three measures develop volume-law scaling (see inset). However, after the mapping to $c$-fermions and conserved charges, we *do* find the phenomenology of the QDL. As shown in Fig. 4.2, at long times we observed area law scaling of $S^c_{\text{PBEE}}$ and volume law scaling of $S^q_{\text{PBEE}}$. Furthermore, since the localization behaviour persists for all system sizes [8], and there is a direct relation between the area-law scaling of $S^{c}_{\mathrm{PBEE}}$ and the localization of fermions, this allows us to infer that this behaviour holds in the thermodynamic limit.

These contrasting results highlight the subtlety of defining a QDL, most crucially on an appropriate choice of the measurement basis. While the dynamics of the $f$ and $c$ fermions is closely related—e.g., all density correlators are the same—they are connected via a non-linear and non-local transformation with a string of spin operators.

### 4.1.3 PBEE Distributions

In the context of our model we found some unexpected results for the projective measure of entanglement used to define the QDL. Particularly, even though we were able to conclusively demonstrate the complete localization of the $f$-fermions in the previous chapters, the PBEE was unable to distinguish this subsystem from that of $\sigma$-spins. This measure is a weighted average, and as such we are not considering all of the information. In this section we investigate the distribution of the terms appearing within the sum (4.1) that may shed some more light on the observed behaviour.

Ideally, we would like to consider the distribution $\sum_{|\phi\rangle} \delta(x - M\mathcal{Z}^{\alpha}_{\phi} S^{\alpha}_{\phi}(t))$, where $M$ is the number of states $|\phi\rangle$ for species $\beta$. However, because of the broadening that is necessary to turn this discrete distribution into a continuous function, dealing with this quantity directly does not give clear results. For this purpose, we consider a more transparent alternative and define the function

$$D^{\alpha}(x,t) = \frac{1}{M} \sum_{|\phi\rangle} \Theta\left(\Delta - |x - M\mathcal{Z}^{\alpha}_{\phi} S^{\alpha}_{\phi}(t)|\right), \tag{4.9}$$

where $x$ takes discrete values, $x = \Delta(2n - 1)$, $n \in \mathbb{N}$, and $\Theta(x)$ is the Heaviside step function. At a given time $t$, this function counts the number of states $|\phi\rangle$ for which the value $M\mathcal{Z}^{\alpha}_{\phi} S^{\alpha}_{\phi}(t)$ is within a window of width $2\Delta$ centred on $x$. In other words, it defines a time dependent histogram, which approximates the distribution we want to consider. We normalize this function such that the sum over all $x$ is 1, and we will refer to it as the distribution of $S^{\alpha}_{\mathrm{PBEE}}$. In Figs. 4.3 and 4.4, we set $\Delta$ such that the plotted range is split into 200 windows, e.g., $\Delta = 0.01$ for Fig. 4.4a. Note that to reduce the noise in these figures that is due to the finite size of our systems, the values are smoothed by short ranged averaging over time, namely, $D^{\alpha}(x,t) \rightarrow \frac{2}{t^2}\int_0^t \mathrm{d}\tau\, \tau D^{\alpha}(x,\tau)$.

In Fig. 4.3 we show the distributions $D^{f}(x,t)$ and $D^{\sigma}(x,t)$, i.e., in the original degrees of freedom. Unlike the averaged value, there is a clear distinction between the localized fermion subsystem and the spin subsystem. In the former we see that the distribution is peaked around the mean value, whereas in the latter we have a very spread out distribution which is peaked at zero. This behaviour is more clearly

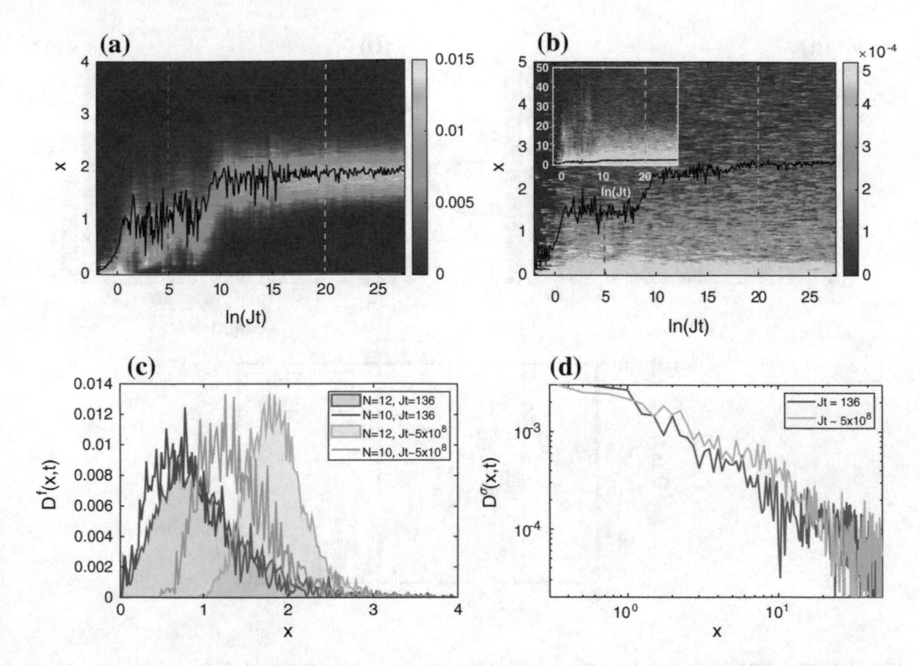


**Fig. 4.3** The PBEE distributions for $h/J = 20$ and $N = 12$. **a** Distribution $D^f(x,t)$ for the $f$-fermions. The black curve indicates the average value, $S^f_{\text{PBEE}}$. **b** the $D^\sigma(x,t)$ distribution, with $S^\sigma_{\text{PBEE}}$ shown in black. (inset) The same data over a larger range of values. **c–d** Cuts in the distribution at $Jt = 136$ and $Jt \sim 5 \times 10^8$, shown in red and yellow. These cuts are indicated in **a** and **b** by dashed lines of the same colour

seen in Fig. 4.3c–d, where we show the distributions for fixed values of time, one at intermediate times $Jt = 136$ and one at long times $Jt \approx 5 \times 10^8$. We see that for the $f$ fermions the distributions are peaked around the mean value but that at long times this mean value increases with systems size, see Fig. 4.3c. On the other hand, for the spins, the distribution decays from zero with approximately power-law behaviour, see Fig. 4.3d, and the volume law scaling is due to the distribution spreading out with more weight at larger values. This can most clearly be seen in the inset of Fig. 4.3b, where the weights are more spread out at longer times than at short and intermediate—after the time-scale set by $(h/J)^2$. We therefore see that the volume law scaling observed in the PBEEs $S^\sigma_{\text{PBEE}}$ and $S^f_{\text{PBEE}}$, have very different origins.

Let us now turn to the $c$ and $q$ degrees of freedom, see Fig. 4.4. Here we also see contrasting behaviour between the $c$ and $q$ degrees of freedom. Furthermore, the behaviour of $D^q(x,t)$ is similar to that of the spins, namely that the distribution is peaked at zero and decays with a power-law. However, there is a stark difference between the distributions for the $c$ and $f$ fermions. In particular, the distribution $D^c(x,t)$ is not peaked around the average value, and we see clear stripes in the distribution, see Fig. 4.4a. These stripes occur between values $(l-1)/2\ln(2)$, with $l \in \mathbb{N}$, and are a consequence of the system splitting into disconnected runs of length $l$,

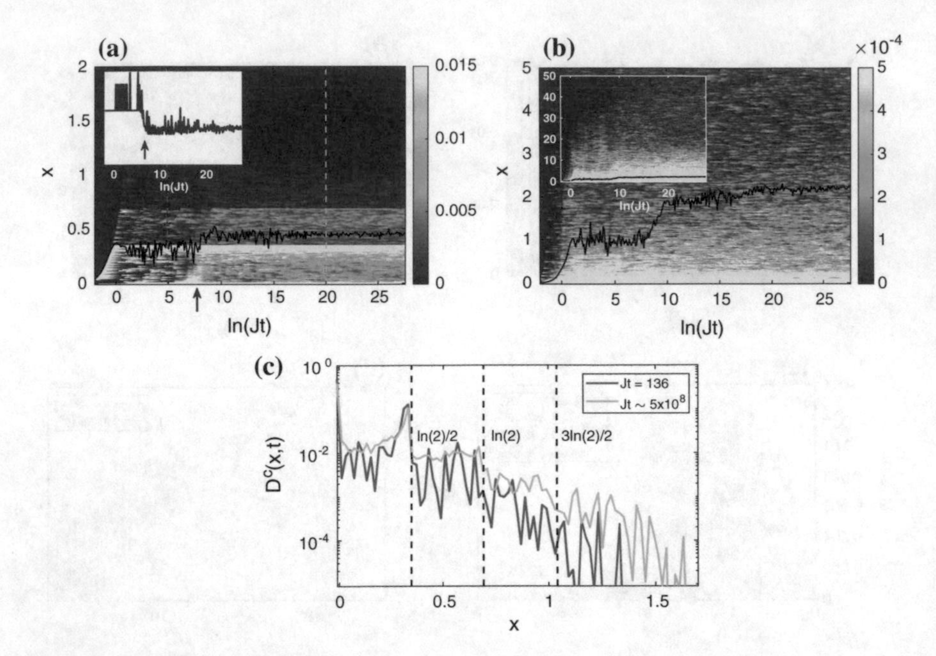


**Fig. 4.4** The PBEE distributions for $h/J = 20$ and $N = 12$. **a** $D^c(x,t)$ distribution with $S^c_{\text{PBEE}}$ shown in black. (inset) The number of values that are in the histogram bin around zero, see Eq. (4.9). The purple arrow on both the main and inset figures indicate the time-scale $Jt \sim (h/J)^2$. **b** Distribution $D^q(x,t)$ for the conserved charges with $S^q_{\text{PBEE}}$ shown in black. (inset) The distribution over a larger range of values. **c** $D^c(x,t)$ for the $c$-fermions at fixed values of $Jt = 136$ and $Jt \sim 5 \times 10^8$. These times are indicated in **a** by dashed lines of the same colour. The black dashed lines in **c** indicate the observed stripes through the distribution at values $(l-1)\ln(2)/2$

see Fig. 4.4c. The entanglement for the $c$ fermions comes from the runs that cross the boundary of the partition. We therefore see overall exponential decay in this distribution reflecting the distribution of run lengths $\sim(1/2)^l$. Most of the weight is at zero due the most probable configuration being runs of length 1. The weight at zero decreases at $Jt \sim (h/J)^2$ and spreads out, as indicated by the arrow in Fig. 4.4a and inset. This is due to the coupling between these runs, which only becomes appreciable on these time scales. The observed behaviour for $D^c(x,t)$ is then most likely due to the 'integrability' of our model and we might expect these stripes to disappear with the introduction of interactions.

## 4.2 Mutual Information

In this section we investigate an alternative measure of entanglement to distinguish the two subsystems—the quantum mutual information. We demonstrate that it is able to distinguish the $f$ and $\sigma$ degrees of freedom, as well as the $c$-fermions from the charges $q$.

### 4.2.1 Quantum Mutual Information

To define the mutual information we split the system into three partitions.[1] We take one of these partitions to be the subsystem corresponding to one of the species, which we label $C$, and we divide the other subsystem into $A$ and $B$. The mutual information between subsystems $A$ and $B$ is then defined as

$$I(A:B) = \frac{1}{2}(S_A + S_B - S_{AB}) \tag{4.10}$$

where $S_A$ is the von Neumann entanglement entropy for the reduced density matrix $\rho_A$, and $S_{AB} = S_{A\cup B} = S_C$. We will use the notation $I_\alpha = I_\alpha(A:B)$ for the mutual information when the $A$ and $B$ partitions correspond to species $\alpha$.

### 4.2.2 Results

The results of the mutual information for our model are shown in Fig. 4.5. Unlike the PBEE we have a clear distinction between $I_f$ and $I_\sigma$, and also between $I_c$ and $I_q$. We observe area law scaling for both $f$ and $c$ fermions and volume law for the spins and the charges.

To understand this behaviour a little further let us separately consider the behaviour of the von Neumann entanglement entropies appearing in the definition Eq. (4.10). These obey the scaling behaviours

$$S^\gamma_{A(B)} = \alpha^\gamma_{A(B)} N + \delta^\gamma_{A(B)}, \qquad S_{AB} = 2\alpha_{AB} N + \delta_{AB}, \tag{4.11}$$

where $N$ is the number of sites, $\gamma \in \{f, \sigma\}$ or $\gamma \in \{c, q\}$, and $S_{AB} = S_C$ is the entanglement between the two species. These entropies consist of a volume law term $\sim N$ and a constant area law term,[2] and we have included a factor of two in the definition of $S_{AB}$ since $|A \cup B| = 2|A|$. If we now consider the combination that defines the mutual information then we find that

$$I_\gamma(A:B) = \frac{1}{2}(\alpha^\gamma_A + \alpha^\gamma_B - 2\alpha_{AB})N + \frac{1}{2}(\delta^\gamma_A + \delta^\gamma_B - \delta_{AB}), \tag{4.12}$$

[1] The mutual information can also be defined for a bipartite mixed state. For a pure state this definition reduces to the standard von Neumann entropy. See Refs. [9, 10] for more details about entanglement measures.

[2] Note that the separation between volume and area law scaling is not clear in our setup. The volume scales with $N$ but so do the boundaries between $A$ and $C$ and between $B$ and $C$. The boundary between $A$ and $B$ on the other hand is constant. Here we simply consider the scaling with $N$ plus a constant term.

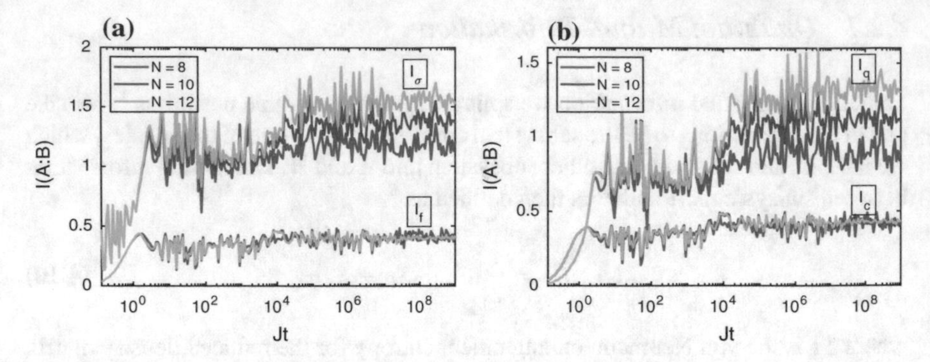


**Fig. 4.5** **a** Mutual information $I_f(A:B)$ and $I_\sigma(A:B)$ for the original $f$ and $\sigma$ degrees of freedom with $h/J = 20$. **b** $I_c(A:B)$ and $I_q(A:B)$ for the $c$ fermions and conserved charges $q$. Data is shown for three different system sizes $N = 8, 10, 12$, which demonstrates area law scaling for $I_f$ and $I_c$ and volume law scaling for $I_\sigma$ and $I_q$

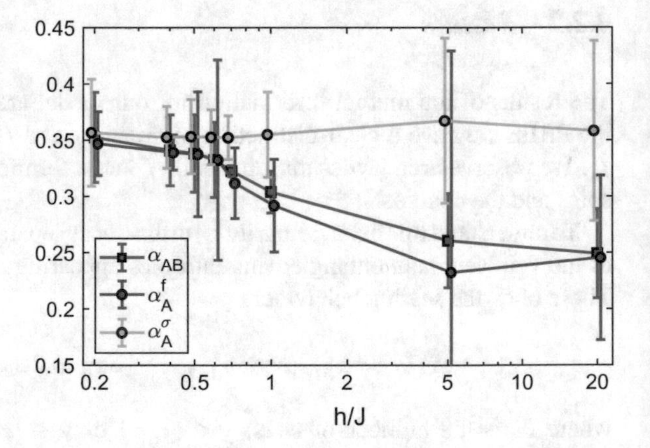


**Fig. 4.6** The coefficients for scaling with systems size $N$ of the von Neumann entanglement entropies $S_{AB}$ and $S_A$ (see Eq. (4.11)) for both the fermions $f$ and the spins $\sigma$ that appear in the definitions of the mutual information $I_f$ and $I_\sigma$. Data is slightly horizontally offset to make the data points and error bars more clearly visible

and in particular we only have volume law scaling if there is an incomplete cancellation in the prefactors. Note that if we had a single species system then we would generally expect $\alpha_A = \alpha_B = \alpha_C$ resulting in area law scaling of the mutual information, which is consistent with Ref. [11] for example.

In Fig. 4.6 we consider the behaviour of the $\alpha$-parameters in the von Neumann entanglement entropies, defined in Eq. (4.11). For the $f$ species we find that $\alpha^f_A (= \alpha^f_B)$ and $\alpha_{AB}$ depend on *h*/*J*—that is, on the localization length—in exactly the same way, resulting in complete cancellation of the volume law prefactor. For the spins on the other hand, $\alpha^\sigma_A$ is independent of *h*/*J* and the prefactors do not cancel exactly, resulting in the volume law scaling seen in Fig. 4.5.

## 4.3 Discussion

Using the projective measures of entanglement defined in Ref. [1] we revealed the behaviour of a quantum disentangled liquid in the state of our system at long times after a quantum quench. However, this diagnostic only gave a positive results in the basis of $c$-fermions and the conserved charges $q$. It was unable to distinguish between the localized $f$-fermions and the spins $\sigma$, both of which exhibited volume law scaling of the PBEE. This indicates the sensitivity of this diagnostic on an appropriate choice of basis, despite the clear identification of localization physics using local measurements investigated in previous chapters.

Rather than simply looking at the PBEE we considered the distribution of the quantities that appear in its definition. These distributions were able to show a distinction between the different degrees of freedom, even when the mean value was unable to. This suggests that this distribution may be a better diagnostic, but since our results were only restricted to a single model and we do not yet have a complete explanation of the results we observe, we are unable to say what should generally be expected. In particular, we found behaviour of the $c$-fermions that is directly related to the presence of conserved charges in our model, and so would not be expected in more general situations. We propose that further investigation is warranted and suggest the interacting generalizations of our model, as well as the heavy-light particle mixture models [2, 3, 12] as ideal testing grounds.

We also considered using the mutual information as an alternate diagnostic tool for the QDL. This measure had some desirable properties and was able to distinguish the localized from the delocalized degrees of freedom in both choices of basis. The volume law scaling for the spins and the charges was explained by an incomplete cancellation of the volume laws of the von Neumann entropies appearing in the definition. While we confirmed this was due to the different dependences of the volume law coefficients on the localization length, we do not yet have a complete explanation for this behaviour. Turning this into a viable measure of a QDL is left for future work.

As a closing remark we would suggest that as well as investigating further the properties of these two measures of entanglement, it would be instructive to consider others. An example that has different properties to both the von Neumann entropy and the mutual information is the negativity [9, 13, 14]. It is clear that despite the simplicity and solubility of our model we can observe non-trivial dynamics and entanglement properties.

# References


1. Grover T, Fisher MPA (2014) Quantum disentangled liquids. J Stat Mech Theory Exp 2014:P10010. https://doi.org/10.1088/1742-5468/2014/10/P10010
2. Schiulaz M, Müller M (2014) Ideal quantum glass transitions: many-body localization without quenched disorder. In: AIP Conference Proceedings, vol 1610, pp 11–23. https://doi.org/10.1063/1.4893505
3. Yao NY, Laumann CR, Cirac JI, Lukin MD, Moore JE (2016) Quasi-many-body localization in translation-invariant systems. Phys Rev Lett 117:240601. https://doi.org/10.1103/PhysRevLett.117.240601
4. Veness T, Essler FHL, Fisher MPA (2017a) Quantum disentangled liquid in the half-filled Hubbard model. Phys Rev B 96:1–9. https://doi.org/10.1103/PhysRevB.96.195153
5. Veness T, Essler FHL, Fisher MPA (2017) Atypical energy eigenstates in the Hubbard chain and quantum disentangled liquids. Philos Trans R Soc A Math Phys Eng Sci 375:20160433. https://doi.org/10.1098/rsta.2016.0433
6. Garrison JR, Mishmash RV, Fisher MPA (2017) Partial breakdown of quantum thermalization in a Hubbard-like model. Phys Rev B 95:054204. https://doi.org/10.1103/PhysRevB.95.054204
7. Smith A, Knolle J, Moessner R, Kovrizhin DL (2017) Absence of ergodicity without quenched disorder: from quantum disentangled liquids to many-body localization. Phys Rev Lett 119:176601. https://doi.org/10.1103/PhysRevLett.119.176601
8. Smith A, Knolle J, Kovrizhin DL, Moessner R (2017) Disorder-free localization. Phys Rev Lett 118:266601. https://doi.org/10.1103/PhysRevLett.118.266601
9. Eisert J (2001) Entanglement in quantum information theory, PhD thesis
10. Wilde MM (2017) Quantum information theory, 2nd edn. Cambridge University Press, Cambridge, UK. https://doi.org/10.1017/9781316809976
11. Wolf MM, Verstraete F, Hastings MB, Cirac JI (2008) Area Laws in quantum systems: mutual information and correlations. Phys Rev Lett 100:070502. https://doi.org/10.1103/PhysRevLett.100.070502
12. Schiulaz M, Silva A, Müller M (2015) Dynamics in many-body localized quantum systems without disorder. Phys Rev B 91:184202. https://doi.org/10.1103/PhysRevB.91.184202
13. Życzkowski K, Horodecki P, Sanpera A, Lewenstein M (1998) Volume of the set of separable states. Phys Rev A 58:883–892. https://doi.org/10.1103/PhysRevA.58.883
14. Vidal G, Werner RF (2002) Computable measure of entanglement. Phys Rev A 65:032314. https://doi.org/10.1103/PhysRevA.65.032314

# Chapter 5
# Out-of-Time-Ordered Correlators

In this chapter we study the out-of-time-ordered correlators (OTOCs) for the gauge field in our model. While originally introduced in the context of quasiclassical approaches to quantum systems [1], OTOCs have recently received renewed interest due to their connections with the emergence of quantum chaotic behaviour [2–4]. This has generated a flurry of activity on OTOCs in the studies of entanglement and information scrambling in integrable [5–7], thermalizing [4] and many-body-localized (MBL) systems [8–10]. OTOCs have been calculated for a number of important models including the transverse field Ising model [5], Luttinger-liquids [6], and random unitary circuits [11–14]. Beyond these examples, the calculation of OTOCs is a difficult task, but nonetheless, some generic features of OTOCs are emerging.

OTOCs are defined as

$$C(t) = \frac{1}{2}\langle |[\hat{A}(t), \hat{B}]|^2 \rangle, \tag{5.1}$$

where $\hat{A}$ and $\hat{B}$ are two operators in the Heisenberg representation, and $[\cdot, \cdot]$ is the commutator. The expectation value $\langle \cdot \rangle$ can in general be taken in one of three ways: (i) full trace, corresponding to an infinite temperature thermal expectation value; (ii) finite temperature expectation value; (iii) with respect to a pure quantum state. In this chapter we will only consider the latter, which corresponds to a global quantum quench protocol, as we have been considering throughout this thesis.

Recently, a motivation for considering OTOCs has been the connection to chaos. These correlators are a quantum analogue of the classical Poisson bracket. In the classical chaotic setting, the Poisson bracket of the position and momentum $\{x(t), p\} = \frac{\partial x(t)}{\partial x(0)}$ grows exponentially in time $\sim e^{\lambda_L t}$, with Lyapunov exponent $\lambda_L$. This exponential growth quantifies the sensitivity to perturbations in chaotic systems and the exponential growth of the OTOCs has been considered as a signature of quantum chaotic behaviour. The Lyapunov exponent for OTOC growth has been shown to be bounded by temperature $\lambda_L \leq 2\pi/(\beta\hbar)$ [3], and the bound is saturated in the Sachdev-Ye-Kitaev model [15, 16]. While this growth behaviour has been estab-


A. Smith, *Disorder-Free Localization*, Springer Theses,
https://doi.org/10.1007/978-3-030-20851-6_5

lished in the large-$N$ and semiclassical limits [17], it is not clear how this applies more generally. In particular, its observation is reliant on a hierarchy of time scales that is absent in many quantum models, and for instance has been shown not to apply to certain integrable models [5–7]. An alternative has been recently suggested in Ref. [18], where they consider the growth along light-rays, that is along lines $|x| = vt$ for different values of $v$, to define a velocity dependent Lyapunov exponent $\lambda_L(v)$.

More generally, OTOCs quantify the spreading of operators and 'operator scrambling' [4, 19]. If $\hat{A}$ and $\hat{B}$ are spatially separated local operators then initially their commutator is zero. As the operator $\hat{A}$ is evolved in time and its support spreads, this commutator will grow. For local operators, the infinite temperature OTOC was shown to be bounded and to display a light-cone causality structure by Lieb and Robinson [20]. The Lieb-Robinson bound can be written as

$$\mathrm{Tr}\left(\|[\hat{A}(t), \hat{B}]\|^2\right) \leq c\, e^{-\frac{L-vt}{\xi}}, \tag{5.2}$$

where $L$ is the separation between the local operators $\hat{A}$ and $\hat{B}$, and $c, v, \xi$ are the constant prefactor, speed and length scale, which are to be determined. In Eq. (1.5) we used a corollary to bound correlators between local operators.[1] Depending on the system, we can also have different behaviours within this light cone. In an integrable system—such as the transverse field Ising model considered in Ref. [5]—the OTOC of local operators is only non-zero close to the light-cone and decays back to zero once it has passed. This is analogous to the propagation of a coherent wavepacket. However, in generic, and particularly thermalizing systems, the OTOC saturates to a non-zero value within the light-cone and the operator $\hat{A}(t)$ is said to be scrambled under time-evolution [4].

OTOCs are generally difficult quantities to deal with, both analytically and to simulate numerically, because of the forward and backward time evolution. Our model provides an ideal setting to study OTOCs due to our mapping to free fermions, which allows us to efficiently simulate these quantities numerically. Notable settings where analytic progress has been made include Luttinger-liquids [6], the integrable transverse field Ising model [5], and random unitary circuit models [11–14]. In the latter an effective hydrodynamic description emerges, which is hypothesised to carry over to thermalizing quantum systems. In localized and MBL systems some numerical progress has also been made using exact diagonalization methods [8, 9] for systems of $L = 11, 12$ sites, which is sufficiently large compared to the localization lengths for the disorder strengths considered. In Ref. [8] the numerical results are also compared with expected behaviour deduced from the $l$-bit representation of the MBL Hamiltonian. The logarithmic spreading of OTOCs observed in these papers is taken as a signature of many-body localization. One of the main results of this chapter is that we demonstrate the logarithmic spreading of OTOCs in an essentially non-interacting and translationally invariant setting.

[1] Note that the constants $c, v, \xi$ for the operator and correlator bounds are not generally the same.

## 5.1 Definition of the OTOC

In this chapter we will consider only an open 1D system and will therefore simplify notation to reduce the number of indices in the remainder of the chapter. In particular we will label the spins on the bonds as $\hat{\sigma}_j^\alpha$ instead of $\hat{\sigma}_{j,j+1}^\alpha$. Given this change of notation we will also allow for site and bond dependent couplings, and write the Hamiltonian as

$$\hat{H} = -\sum_{j=1}^{N-1} J_j \hat{\sigma}_j^z \left(\hat{f}_j^\dagger \hat{f}_{j+1} + \text{H.c.}\right) - \sum_{j=2}^{N-1} h_j \hat{\sigma}_{j-1}^x \hat{\sigma}_j^x, \tag{5.3}$$

where $J_j = J$ and $h_j = h$ are initially taken as constant.

The quantities that we will discuss in this chapter are the out-of-time-ordered correlators for the gauge field

$$C_{jl}^{\alpha\beta}(t) = \frac{1}{2}\langle\Psi|\,|[\hat{\sigma}_j^\alpha(t), \hat{\sigma}_l^\beta]|^2|\Psi\rangle = 1 - \text{Re}\langle\Psi|\hat{\sigma}_j^\alpha(t)\hat{\sigma}_l^\beta\hat{\sigma}_j^\alpha(t)\hat{\sigma}_l^\beta|\Psi\rangle, \tag{5.4}$$

where $\alpha, \beta \in \{x, z\}$, and $|\Psi\rangle$ is a particular initial state for the fermions and spins. The equality follows from the commutation relations of the Pauli matrices. We will also use the shorthand notation

$$F_{jl}^{\alpha\beta}(t) = \langle\Psi|\hat{\sigma}_j^\alpha(t)\hat{\sigma}_l^\beta\hat{\sigma}_j^\alpha(t)\hat{\sigma}_l^\beta|\Psi\rangle, \tag{5.5}$$

for the time-dependent contribution to the OTOC, and so $C_{jl}^{\alpha\beta}(t) = 1 - \text{Re}\, F_{jl}^{\alpha\beta}(t)$. We will consider the initial state to be a tensor product $|\Psi\rangle = |S\rangle \otimes |\psi\rangle$ of the spins in a $z$-polarized state $|S\rangle = |\cdots\uparrow\uparrow\uparrow\cdots\rangle$ and a Fermi-sea Slater determinant for the fermions, i.e., $|\psi\rangle$ to be the half-filled ground state of Hamiltonian $\hat{H}_{\text{FS}} = -\sum_{\langle ij\rangle}\hat{c}_i^\dagger\hat{c}_j$. This Fermi-sea initial state is regularly prepared in cold atom optical lattice experiments [21–23].

### *5.1.1 Mapping to Free Fermions: Double Loschmidt Echo*

Similarly to the correlators discussed in the previous chapters, the OTOCs can be mapped to free-fermion correlators. This allows us to compute these quantities efficiently and also gives us an insight into their properties. Let us consider the non-trivial piece of the OTOC, $F_{jl}^{\alpha\beta}(t)$. We can commute the Pauli operators one by one through the unitary evolution operators so that they act on the initial state to the right. Due to the fact that these operators do not commute with the Hamiltonian, we will modify the Hamiltonians in the process. We can see that $\hat{\sigma}_j^z$ commutes with the hopping term but anti-commutes with two of the Ising coupling terms, $\hat{\sigma}_{j-1}^x\hat{\sigma}_j^x$ and $\hat{\sigma}_j^x\hat{\sigma}_{j+1}^x$. Conversely, $\hat{\sigma}_j^x$ commutes with the Ising coupling terms and anti-commutes with the hopping term. We can therefore write in shorthand notation $\hat{\sigma}_j^\alpha e^{\pm i\hat{H}t} = e^{\pm i\hat{H}_j^\alpha t}\hat{\sigma}_j^\alpha$ and $\hat{\sigma}_j^\alpha\hat{\sigma}_l^\beta e^{\pm i\hat{H}t} = e^{\pm i\hat{H}_{jl}^{\alpha\beta}t}\hat{\sigma}_j^\alpha\hat{\sigma}_l^\beta$, where $\hat{H}_j^\alpha$ has $h_j, h_{j+1} \to -h_j, -h_{j+1}$ on sites $j$ and $j+1$ if $\alpha = z$, and if $\alpha = x$ we have $J_j \to -J_j$ on the bond connecting sites $j$ and $j+1$. Similarly for $\hat{H}_{jl}^{\alpha\beta}$. This allows us to write

$$F_{jl}^{\alpha\beta}(t) = \langle\Psi|e^{i\hat{H}t}e^{-i\hat{H}_j^\alpha t}e^{i\hat{H}_{jl}^{\alpha\beta}t}e^{-i\hat{H}_l^\beta t}\hat{\sigma}_j^\alpha\hat{\sigma}_l^\beta\hat{\sigma}_j^\alpha\hat{\sigma}_l^\beta|\Psi\rangle = \langle\Psi|e^{i\hat{H}t}e^{-i\hat{H}_j^\alpha t}e^{i\hat{H}_{jl}^{\alpha\beta}t}e^{-i\hat{H}_l^\beta t}|\Psi\rangle, \tag{5.6}$$

where in the second equality we have used the commutation relations for the Pauli matrices and that $(\hat{\sigma}_j^\alpha)^2 = 1$. To clarify the shorthand notation used in this expression, let us for concreteness consider $\alpha = x$, $\beta = z$, then $\hat{H}_j^\alpha$ has $J_j \to -J_j$, $\hat{H}_{jl}^{\alpha\beta}$ has $J_j, h_l, h_{l+1} \to -J_j, -h_l, -h_{l+1}$, and $\hat{H}_l^\beta$ has $h_l, h_{l+1} \to -h_l, -h_{l+1}$, all relative to $\hat{H}$.

We can use the duality transformation to write Eq. (5.6) as a free-fermion correlator, namely

$$F_{jl}^{\alpha\beta}(t) = \frac{1}{2^{N-2}}\sum_{\{q_i\}=\pm 1}\langle\psi|e^{i\hat{H}(q)t}e^{-i\hat{H}_j^\alpha(q)t}e^{i\hat{H}_{jl}^{\alpha\beta}(q)t}e^{-i\hat{H}_l^\beta(q)t}|\psi\rangle, \tag{5.7}$$

where

$$\hat{H}(q) = -\sum_{j=1}^{N-1} J_j\left(\hat{c}_j^\dagger\hat{c}_{j+1} + \text{H.c.}\right) + \sum_{j=2}^{N-1} h_j q_j(2\hat{c}_j^\dagger\hat{c}_j - 1), \tag{5.8}$$

with $q_j = \pm 1$ now as classical variables as in Eq. (2.22). See Sect. 2.2 for more details on this transformation. The OTOC therefore takes the form of a *double Loschmidt echo* for free-fermions, averaged over binary disorder, and we compute the correlators appearing in this sum using determinants of matrices, as explained in Appendix B.

### 5.1.2 Interpretation of the Double Loschmidt Echo

The time dependent part of the OTOC

$$F_{jl}^{\alpha\beta}(t) = \langle\Psi|e^{i\hat{H}t}e^{-i\hat{H}_j^\alpha t}e^{i\hat{H}_{jl}^{\alpha\beta}t}e^{-i\hat{H}_l^\beta t}|\Psi\rangle. \tag{5.9}$$

is in the form of a non-standard double Loschmidt echo. Here we briefly offer an interpretation of this quantity and suggest that it should be a useful quantity for studying correlation spreading in quantum systems more generally. Let us for this purpose talk on a more abstract and general level and consider $\hat{H}_j^\alpha = \hat{H} + \hat{V}_j^\alpha$, and $\hat{H}_{jl}^{\alpha\beta} = \hat{H} + \hat{V}_j^\alpha + \hat{V}_l^\beta$, where $\hat{V}_j^\alpha$ and $\hat{V}_j^\beta$ are two types of local perturbations labelled by $\alpha$ and $\beta$ and localized on site $j$.

As a starting point, let us consider the two cases where one of the local perturbations is zero. First, if $\hat{V}_j^{\alpha} = 0$, then we have that

$$F_{jl}^{\alpha\beta}(t) = \langle\Psi|e^{i\hat{H}t}e^{-i\hat{H}t}e^{i\hat{H}_l^{\beta}t}e^{-i\hat{H}_l^{\beta}t}|\Psi\rangle = 1, \tag{5.10}$$

since each time evolution is followed by its exact time-reversal. Secondly, if instead $\hat{V}_l^{\beta} = 0$, then

$$F_{jl}^{\alpha\beta}(t) = |e^{i\hat{H}_j^{\alpha}t}e^{-i\hat{H}t}|\Psi\rangle|^2 = 1, \tag{5.11}$$

which also follows from the unitarity of the Hamiltonian evolution. A non-trivial response is therefore the result of the mutual influence of the two perturbations. Put another way, the response can only be non-trivial if the influence of the local perturbation at site $j$ in

$$e^{i\hat{H}_j^{\alpha}t}e^{-i\hat{H}t}|\Psi\rangle, \quad \text{and} \quad e^{i\hat{H}_{jl}^{\alpha\beta}t}e^{-i\hat{H}_l^{\beta}t}|\Psi\rangle, \tag{5.12}$$

differ, due to the presence of the perturbation at site $l$. The double Loschmidt echo therefore quantifies only the mutual influence of the two separated local perturbations and not the effect of either individually. We hence suggest that this is a natural quantity to consider beyond the scope considered in the main text. In particular, if the perturbations $\alpha$ and $\beta$ are the same type, the double Loschmidt echo measures the spatial influence of that type of perturbation as a function of time and provides more information than the Loschmidt echo alone.

Note, in cases where the types of perturbation $\alpha$ and $\beta$ differ, that while swapping the two perturbations in Eq. (5.9) results in an inequivalent quantity, a non-trivial response in one implies it in the other. This is observed in Fig. 5.4—while the two correlators differ in their details, they agree qualitatively.

## 5.2 Results

Here we consider the numerical data for these OTOCs of the gauge field. All results below will be for $N = 100$ sites with open boundary conditions and sampling over 10,000 charge configurations in all figures except Fig. 5.3b where we use 20,000. We fix $j$ to correspond to the central bond and vary $l$. Both $C_r^{\alpha\beta}(t)$ and $F_r^{\alpha\beta}(t)$ are then presented as functions of the separation, $r$, between these two bonds.

### 5.2.1 Short-Time Behaviour

First, we consider the short-time behaviour of the OTOCs. In all cases we observe power-law growth from zero, as shown for a particular correlator in Fig. 5.1. This

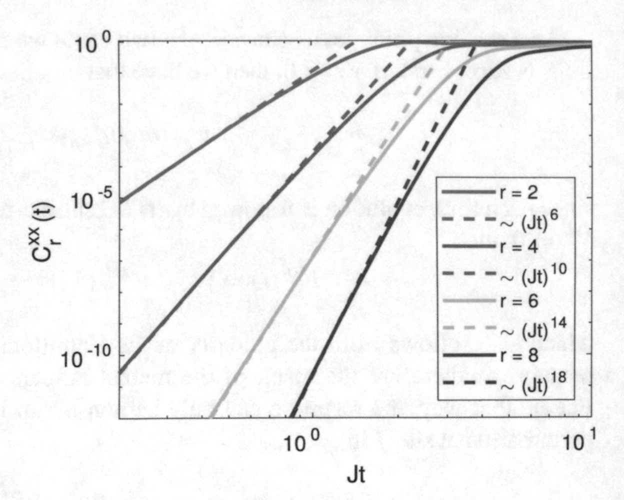


**Fig. 5.1** Short-time behaviour of the OTOCs on a log-log scale where $r$ is the separation between spins in the OTOC. Data shown is for $C_r^{xx}(t)$ with $h = 0.4J$, but we see similar initial power-law growth for all the OTOCs that we consider. The dashed lines indicate fits to the short-time asymptotic behaviour with the form $\sim(Jt)^{2r+2}$

is consistent with the analysis in Refs. [5, 6]. This power-law behaviour can be extracted from the Baker-Cambpell-Haussdorf formula for the time evolution of the operators, namely

$$\hat{\sigma}_j^{\alpha}(t) = \sum_{n=0}^{\infty} \frac{(it)^n}{n!}[\hat{H}, \hat{\sigma}_j^{\alpha}]_n, \tag{5.13}$$

where $[A, B]_n = \left[A, [A, B]_{n-1}\right]$ is the nested commutator. Using the expression in Eq. (5.4), the leading time dependence for the OTOC can then be seen to be $\sim t^{2n}/(n!)^2$, where $n$ is the smallest value for which $\left[[\hat{H}, \hat{\sigma}_j^{\alpha}]_n, \hat{\sigma}_l^{\beta}\right] \neq 0$. Since $\hat{H}$ is a local Hamiltonian, the operator $[\hat{H}, \hat{\sigma}_j^{\alpha}]_n$ must have finite support proportional to $n$, and therefore the lowest-order contribution comes when $n$ is proportional to the separation, $r$, between bonds. This analysis agrees with the short-time behaviour that we observe, see Fig. 5.1. In particular, we find that $C_r^{xx}(t)$ has the asymptotic form $\sim (Jt)^{2r+2}$. Reference [5] suggests that these arguments hold for any OTOCs of bounded local operators with local Hamiltonian evolution.

### 5.2.2 *Exponential Localization of $C_r^{xx}(t)$*

In Figs. 5.2, 5.3 and 5.4 we consider the spreading of correlations in the four distinct OTOCs for the $\sigma$ operators, starting with $C_r^{xx}(t)$ in Fig. 5.2. At short-times, and particularly for small values of the Ising coupling, e.g., $h = 0.4J$ shown in Fig. 5.2a, we find a linear light-cone behaviour (see inset), which agrees with the Lieb-Robinson bound (5.2) with velocity $v = 2J$. At longer times the spreading halts and we find only short-range correlations at long-times. In Fig. 5.2b we show the spreading with $h = 0.8J$ for which the localization length is shorter and the spreading halts more quickly.

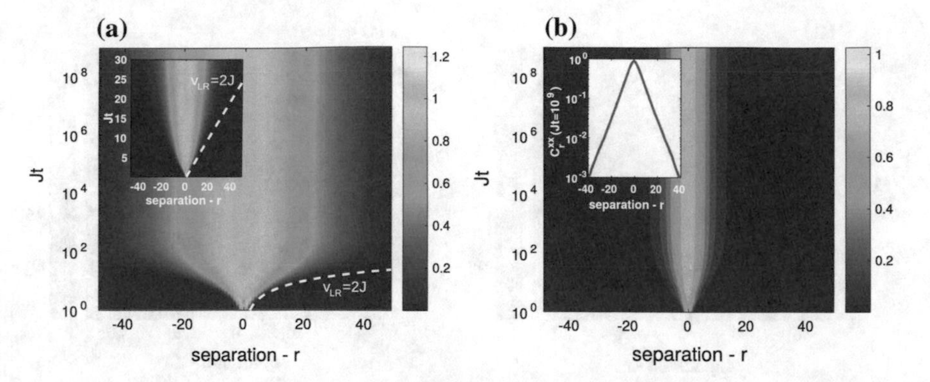


**Fig. 5.2** Correlation spreading in $C_r^{xx}(t)$ starting from a $z$-polarized spin state and fermions in a half-filled Fermi-sea. **a** Data for $h = 0.4J$ with a logarithmic time scale. Dashed line indicates the linear light-cone with Lieb-Robinson velocity $v_{\mathrm{LR}} = 2J$. (inset) the same data but with linear time scale and for shorter times. **b** Results for $h = 0.8J$. (inset) Long time spatial distribution of correlations on a log scale showing exponential tails

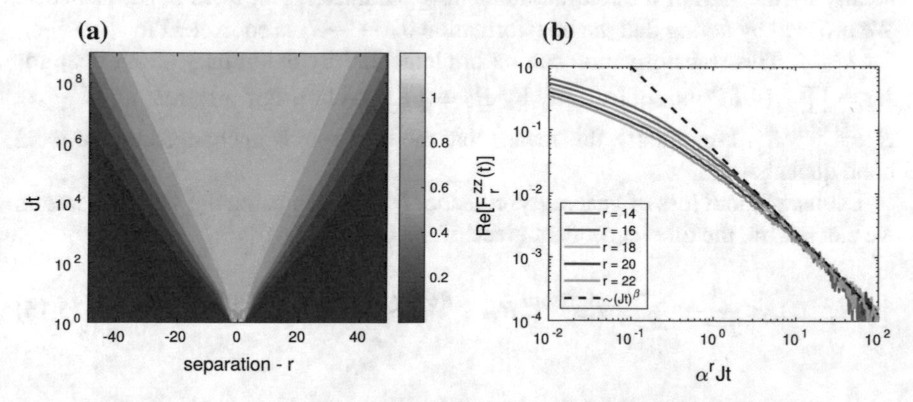


**Fig. 5.3** **a** Data for $C_r^{zz}(t)$ for $h = 0.8J$. **b** Long-time behaviour of $\mathrm{Re}[F_r^{zz}(t)]$, where $r$ is the separation between spins in the OTOC. We rescale the time with $\alpha = 0.5066$, so that the curves coincide at long-times and are can be compared with the power-law $\sim (Jt)^\beta$, with $\beta = -1.3$

The inset shows the spatial correlations at long-times, which decay exponentially with separation. These short range correlations were also observed for OTOCs in an Anderson localized system in Ref. [9].

The time-dependent part of the $C_r^{xx}(t)$ OTOC is

$$F_r^{xx}(t) = \frac{1}{2^{N-2}} \sum_{\{q_i\}=\pm 1} \langle \psi | e^{i\hat{H}(q)t} e^{-i\hat{H}_j^x(q)t} e^{i\hat{H}_{jl}^{xx}(q)t} e^{-i\hat{H}_l^x(q)t} | \psi \rangle, \tag{5.14}$$

where for $\hat{H}_j^x(q)$ we have $J_j \to -J_j$, for $\hat{H}_{jl}^{xx}(q)$ we have $J_j, J_l \to -J_j, -J_l$, and for $\hat{H}_l^x(q)$ we have $J_l \to -J_l$, all relative to $\hat{H}(q)$. These four Hamiltonians differ

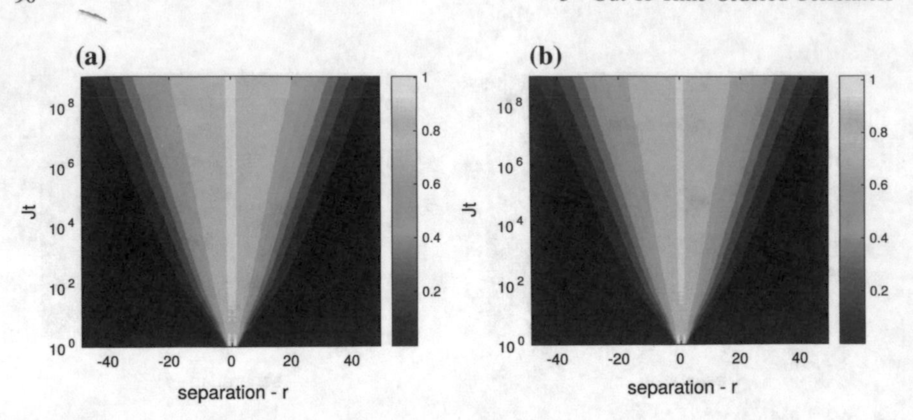


**Fig. 5.4** Spreading of the OTOCs involving different spin components: **a** $C_r^{zx}(t)$; **b** $C_r^{xz}(t)$. Results are shown for $h = 0.8J$ and on a logarithmic scale

locally by the sign of the relevant tunnelling parameter, i.e., local bond quenches. We proceed by noting that the transformation $J_j \to -J_j$ is equivalent to $\hat{c}_i \to -\hat{c}_i$, for $i \leq j$. This transformation can be implemented by the unitary string operator $\hat{R}_j = \prod_{i\leq j}(-1)^{\hat{n}_i}$ or equivalently by $\hat{R}'_j = \prod_{i>j}(-1)^{\hat{n}_i}$. For instance, $e^{\pm i\hat{H}_j^x(q)t} = \hat{R}_j e^{\pm i\hat{H}(q)t}\hat{R}_j$. Importantly, this means that the spectrum is unchanged by the local bond quenches.

Let us, without loss of generality, consider $l > j$, then using the string operators we can rewrite the time-dependent piece of the OTOC (5.6) as

$$F_r^{xx}(t) = \frac{1}{2^{N-2}} \sum_{\{q_j\}\pm 1} \langle\psi|e^{i\hat{H}(q)t}\hat{R}_j e^{-i\hat{H}(q)t}\hat{R}'_l e^{i\hat{H}(q)t}\hat{R}_j e^{-i\hat{H}(q)t}\hat{R}'_l|\psi\rangle, \quad (5.15)$$

where we have used the facts that $[\hat{R}_j, \hat{R}'_l] = 0$ and $(\hat{R}_j)^2 = 1$. The further means that the OTOC can be written as

$$C_r^{xx}(t) = \frac{1}{2^{N-1}} \sum_{\{q_j\}\pm 1} \langle\psi||[\hat{R}_j(t), \hat{R}'_l]|^2|\psi\rangle, \quad (5.16)$$

where $\hat{R}_j(t) = e^{i\hat{H}(q)t}\hat{R}_j e^{-i\hat{H}(q)t}$, which implicitly depend on the charge configuration. Note the extra prefactor of $1/2$ coming from the definition of the OTOC in Eq. (5.4). This OTOC is therefore reduced to a disorder-averaged OTOC of parity operators for the free fermion Hamiltonian (5.8). We can understand the observed behaviour by considering the operator commutator for random configurations of the potential, i.e., for the individual terms of the sum in Eq. (5.16).

The operator $\hat{R}_j$ measures the fermion parity on the chain of sites to the left of $j$, and similarly $\hat{R}'_l$ measures the parity to the right of site $l$. Therefore, to get $[\hat{R}_j(t), \hat{R}'_l] \neq 0$ would require particle transport between the left of site $j$ and the right of site $l$ due to the time evolution. However, since for a typical charge configuration we have a disordered potential, the fermions will be localized meaning that there is an exponentially small probability for the required particle transport. This implies that $\langle\psi||[\hat{R}_j(t), \hat{R}_l']|^2|\psi\rangle \propto e^{-|j-k|/\zeta}$ as $t \to \infty$, where $\zeta$ is a length scale that is a function of the single-particle localization lengths. This long-time behaviour is observed in Fig. 5.2b, where we find exponential tails for the spatial distribution of correlations.

### 5.2.3 *Logarithmic Spreading of $C_r^{zz}(t)$*

Next we look at the $C_r^{zz}(t)$ correlator for $h = 0.8J$ in Fig. 5.3. In particular we observe logarithmic spreading of correlations, which contrasts $C_r^{xx}(t)$ and can be seen in Fig. 5.3a, which has linear contours on a logarithmic scale. Note that for $h = 0.8J$ the localization length for the fermions is $\lambda \approx 2.15$, which is much less than the system size and the scale of correlation spreading.

Whereas $C_r^{xx}(t)$ corresponded to a disorder-averaged double Loschmidt echo procedure with local bond quenches, the quenches in the $C_r^{zz}(t)$ OTOC are of the local potential. Importantly, this type of quench does not preserve the spectrum. The time-dependent part of the $C_r^{zz}(t)$ OTOC is

$$F_r^{zz}(t) = \frac{1}{2^{N-2}} \sum_{\{q_i\}=\pm 1} \langle\psi|e^{i\hat{H}(q)t} e^{-i\hat{H}_j^z(q)t} e^{i\hat{H}_{jl}^{zz}(q)t} e^{-i\hat{H}_l^z(q)t}|\psi\rangle, \tag{5.17}$$

where for $\hat{H}_j^z(q)$ we have $h_j, h_{j+1} \to -h_j, -h_{j+1}$, for $\hat{H}_{jl}^{zz}(q)$ we have $h_j, h_{j+1}, h_l, h_{l+1} \to -h_j, -h_{j+1}, -h_l, -h_{l+1}$, and for $\hat{H}_l^z(q)$ we have $h_l, h_{l+1} \to -h_l, h_{l+1}$, all relative to $\hat{H}(q)$. This correlator corresponds to a double Loschmidt echo procedure with local density quenches of the form $\hat{V} = \pm 2h(\hat{n}_j \pm \hat{n}_{j+1})$, where the signs depend on the particular charge configuration.

We can understand the logarithmic spreading seen in Fig. 5.3a using perturbative arguments similar to those in Ref. [24] for the long-time behaviour of the standard Loschmidt echo. More explicitly, let us consider the single-particle correlators

$$\mathcal{L}_r^{zz}(t) = \langle\psi|e^{i\hat{H}(q)t} e^{-i\hat{H}_j^z(q)t} e^{i\hat{H}_{jl}^{zz}(q)t} e^{-i\hat{H}_l^z(q)t}|\psi\rangle, \tag{5.18}$$

for a typical charge configuration, which are averaged over in Eq. (5.6). We can use the Lehmann decomposition to express it in terms of the many-body eigenstates and energies. This gives

$$\mathcal{L}_r^{zz}(t) = \sum_{\lambda,\mu_j,\nu_{jl},\chi_l} \langle\psi|\lambda\rangle\langle\lambda|\mu_j\rangle\langle\mu_j|\nu_{jl}\rangle\langle\nu_{jl}|\chi_l\rangle\langle\chi_l|\psi\rangle e^{i\Delta E t}, \tag{5.19}$$

where $\Delta E = E_\lambda - E_{\mu_j} + E_{\nu_{jl}} - E_{\chi_l}$. In this expansion, the states $|\lambda\rangle, |\mu_j\rangle, |\nu_{jl}\rangle, |\chi_l\rangle$ are many-body eigenstates of the Hamiltonians $\hat{H}(q), \hat{H}_j^z(q), \hat{H}_{jl}^{zz}(q), \hat{H}_l^z(q)$, respectively, and $E_\lambda, E_{\mu_j}, E_{\nu_{jl}}, E_{\chi_l}$ are the corresponding energy eigenvalues.

We then proceed by first making the approximation that the wavefunctions are only locally perturbed and, in particular, we assume that $\langle\lambda|\mu_j\rangle \approx \delta_{\lambda,\mu_j}$, and similarly for all the eigenstate overlaps appearing in Eq. (5.19), and we neglect the modification of the eigenvectors due to the perturbation. This is justified by the fact that the single-particle eigenstates are localized and the local potential perturbation only locally changes the eigenstates, and in particular the exponential profile is preserved.

With this approximation taken into account, the expression (5.19) reduces to

$$\mathcal{L}_r^{zz}(t) \approx \sum_\lambda \langle\psi|\lambda\rangle\langle\lambda|\psi\rangle e^{i\Delta E(\lambda)t}, \tag{5.20}$$

where $\Delta E(\lambda) = E_\lambda - E_{\lambda_j} + E_{\lambda_{jl}} - E_{\lambda_l}$. The deviation of $\mathcal{L}_r^{zz}(t)$ from 1 is thus determined by $\Delta E(\lambda)$, wherein the energies $E_{\lambda_j}, E_{\lambda_{jl}}, E_{\lambda_l}$ are the perturbed energies corresponding to the same eigenstate $|\lambda\rangle$. We estimate this energy difference using a second-order perturbation expansion.

For convenience, let us state the result of perturbation theory for the eigenergies up to second order (see, e.g., Ref. [25]). Consider a Hamiltonian $\hat{H}$ and a perturbation $\epsilon\hat{V}$. In our case, $\hat{V}$ takes the form $\hat{V}_j = \pm\hat{n}_j \pm \hat{n}_{j+1}$, $\hat{V}_{jl} = \pm\hat{n}_j \pm \hat{n}_{j+1} \pm \hat{n}_l \pm \hat{n}_{l+1}$, or $\hat{V}_l = \pm\hat{n}_l \pm \hat{n}_{l+1}$, with $\epsilon = 2h$. The energies of the perturbed Hamiltonian $\hat{H} + \hat{V}$ are then given by

$$E_\lambda = E_\lambda^{(0)} + \epsilon\langle\lambda|\hat{V}|\lambda\rangle + \epsilon^2 \sum_{\mu\neq\lambda} \frac{|\langle\lambda|\hat{V}|\mu\rangle|^2}{E_\lambda^{(0)} - E_\mu^{(0)}} + \mathcal{O}(\epsilon^3), \tag{5.21}$$

where $|\lambda\rangle, |\mu\rangle$ are unperturbed eigenstates, and $E_\lambda^{(0)}, E_\mu^{(0)}$ are the unperturbed energy eigenvalues. The first-order corrections to the energies cancel in the energy difference $\Delta E(\lambda)$ and the leading-order correction from second-order is

$$\Delta E(\lambda) \approx \pm 8h^2 \mathrm{Re} \sum_{\mu\neq\lambda} \frac{\langle\lambda|\hat{n}_j \pm \hat{n}_{j+1}|\mu\rangle\langle\mu|\hat{n}_l \pm \hat{n}_{l+1}|\lambda\rangle}{E_\lambda^{(0)} - E_\mu^{(0)}}, \tag{5.22}$$

where the signs are determined by the particular charge configuration. Since for a typical disorder configuration the system is Anderson localized, the energy eigenvalues are uncorrelated and the difference $E_\lambda^{(0)} - E_\mu^{(0)}$ is a random function of the states $|\lambda\rangle, |\mu\rangle$. The operator

$$\sum_{\mu\neq\lambda} \frac{|\mu\rangle\langle\mu|}{E_\lambda^{(0)} - E_\mu^{(0)}}, \tag{5.23}$$

is therefore a random diagonal operator in the basis of eigenstates. Furthermore, for a given localized eigenstate $|\lambda\rangle$, the density correlations in Eq. (5.22) are exponentially decaying with the separation $r$. The energy difference to second-order will therefore take the functional form

$$\Delta E(\lambda) \sim \pm 8h^2 C(\{E_\mu\}, |\lambda\rangle)\, e^{-r/\xi}, \quad (5.24)$$

where $C(\{E_\mu\}, |\lambda\rangle)$ is a random function of the energies of the many-body states of $\hat{H}(q)$ with dimensions of inverse energy, $r$ is the separation between the bonds $j$ and $l$, and the length scale $\xi$ is a function of the single-particle localization lengths. Plugging this back into Eq. (5.20) we find that the correlator deviates from 1 when $h^2 C e^{-r/\xi} t \sim 1$. Rearranging this expression we find the relationship

$$r \sim \ln\left(\frac{h^2 C t}{\xi}\right), \quad (5.25)$$

which defines the light-cone for the OTOC. Averaging over $|\lambda\rangle$ and the charge configurations $\{q_j\}$ results in a logarithmic light-cone, as observed in Fig. 5.3a.

While these arguments are perturbative, we also see this logarithmic light-cone beyond the perturbative regime. We argue that they can be extended due to the decreasing localization length with increasing $h$. This means that the orthogonality assumptions, e.g., $\langle\lambda|\mu_j\rangle \approx \delta_{\lambda,\mu_j}$, remain valid. The energy corrections to all orders are also correlators of density operators and random diagonal operators and therefore decay exponentially with the separation $r$.

At long times the $C_t^{zz}(t)$ OTOC has power-law behaviour, as can be seen in Fig. 5.3b, which shows the time-dependent piece $F_r^{zz}(t)$. The exponent appears to be approximately independent of the separation as we find by scaling the time by $Jt \to \alpha^r Jt$ such that the curves coincide with each other. The value $\alpha = 0.5066$ and the exponent $\beta = -1.3$ of the power-law are found empirically. The authors of Refs. [24, 26] also found power-law decay of the Loschmidt echo for localized systems. Similar power-law decay was also observed in the context of OTOCs in a many-body localized system [27], the transverse-field Ising model [5], as well as for the XY spin-chain and symmetric Kitaev chain in Ref. [7]. Our results for the logarithmic spreading and the power-law decay of $C_r^{zz}(t)$ are also consistent with the analysis in Ref. [24] that reveals a power-law decay of the standard Loschmidt echo in a localized system.

### 5.2.4 Spreading of Mixed OTOCs

Finally, we consider two inequivalent OTOCs involving different spin components $\hat{\sigma}^z$ and $\hat{\sigma}^x$, namely $C_r^{zx}(t)$ and $C_r^{xz}(t)$, see Fig. 5.4. Note that in our definition, the first subscript corresponds to the fixed, bond, which is measured at time $t$, whereas the second subscript is the bond that is varied and measured at time $t = 0$. These correlators show qualitatively similar behaviour. For short separations we find nearly time independent contours signifying localization behaviour. However, for larger separations we observe additional spreading of correlations. While the analytic arguments presented above do not apply to this case of mixed-component correlators, the logarithmic spreading appears to be a more general feature. This behaviour is consistent with the interpretation of the double Loschmidt echo that we presented in Sect. 5.1.2.

## 5.3 Discussion

We have explored the behaviour of the four distinct OTOCs for the gauge field in our model, where we revealed a rich phenomenology beyond that which has been observed for Anderson-localized systems. In our results we do not see the exponential growth of the OTOC that has been attributed to chaotic behaviour. Instead we find power-law growth consistent with that found for the transverse-field Ising model [5] and Luttinger-liquids [6]. While our model does not contain the ingredients of many-body localization, we find correlation spreading that is logarithmic, and in some cases the model shows a complete lack of correlation spreading.

We argue that the logarithmic spreading of correlations is due to the disorder-averaged double Loschmidt echo form of the correlator in terms of free-fermions, in contrast to the standard commutators of fermion creation/annhilation operators [4, 8, 9]. When the local perturbations in the double Loschmidt echo correlator change the energy spectrum we get logarithmic spreading. Our results are also consistent with those in Ref. [24] for the standard Loschmidt in a disordered system. We note that a similar slow spreading of free-fermion OTOCs, as well as logarithmic entanglement growth, has been observed in Ref. [7] due to a different mechanism, namely a non-local form of the operators in the computational basis. Logarithmic entanglement growth has also been in observed in the fine-tuned critical phase of a non-interacting central-site model [28]. In cases where the effect of changes of bond signs can be gauged away, we find that the correlations are exponentially localized, as is also observed for standard fermion correlators in quenched disorder models of Anderson localization, see e.g. Ref. [9]. Although these quantities arise as natural spin OTOCs in our model, we propose that the resulting double Loschmidt echo form of the correlators may be useful in the studies of correlation spreading more generally and provides more information than the standard Loschmidt echo alone.

Our model provides an ideal setting for further study of OTOCs. Due to the mapping to free fermions we are able to access large systems in numerical simulations, which can also be performed in the case of thermal or infinite temperature expectation values. We leave a full investigation of the initial state and temperature dependence of the OTOC for future work. Away from this free-fermion limit we can make connection to the physics of MBL, spin confinement, the Falicov-Kimball model and the Hubbard model, as discussed in previous chapters. Unfortunately, in these cases we are faced with severe limitations on numerically-accessible system sizes and time scales. Excitingly, there is also the prospect of simulating OTOCs in experiments [29, 30] and we also contribute a proposal for an experimental protocol in Chap. 7.

## References

1. Larkin AI, Ovchinnikov YN (1969) Quasiclassical method in the theory of superconductivity. Sov Phys JETP 28:1200–1205
2. Berry M (1989) Quantum chaology, not quantum chaos. Phys Scr 40:335–336. https://doi.org/10.1088/0031-8949/40/3/013
3. Maldacena J, Shenker SH, Stanford D (2016) A bound on chaos. J High Energy Phys 2016:106. https://doi.org/10.1007/JHEP08(2016)106
4. Bohrdt A, Mendl CB, Endres M, Knap M (2017) Scrambling and thermalization in a diffusive quantum many-body system. New J Phys 19:063001. https://doi.org/10.1088/1367-2630/aa719b
5. Lin C-J, Motrunich OI (2018) Out-of-time-ordered correlators in a quantum Ising chain. Phys Rev B 97:144304. https://doi.org/10.1103/PhysRevB.97.144304
6. Dóra B, Moessner R (2017) Out-of-time-ordered density correlators in luttinger liquids. Phys Rev Lett 119:026802. https://doi.org/10.1103/PhysRevLett.119.026802
7. McGinley M, Nunnenkamp A, Knolle J (2018) Slow growth of entanglement and out-of-time-order correlators in integrable disordered systems, 1–10. arXiv:1807.06039
8. Chen X, Zhou T, Huse DA, Fradkin E (2017) Out-of-time-order correlations in many-body localized and thermal phases. Ann Phys 529:1600332. https://doi.org/10.1002/andp.201600332
9. Huang Y, Zhang YL, Chen X (2017) Out-of-time-ordered correlators in many-body localized systems. Ann Phys 529:1–6. https://doi.org/10.1002/andp.201600318
10. Fan R, Zhang P, Shen H, Zhai H (2017) Out-of-time-order correlation for many-body localization. Sci Bull 62:707–711. https://doi.org/10.1016/j.scib.2017.04.011
11. Nahum A, Vijay S, Haah J (2018) Operator Spreading in Random Unitary Circuits. Phys Rev X 8:021014. https://doi.org/10.1103/PhysRevX.8.021014
12. von Keyserlingk CW, Rakovszky T, Pollmann F, Sondhi SL (2018) Operator hydrodynamics, OTOCs, and entanglement growth in systems without conservation laws. Phys Rev X 8:021013. https://doi.org/10.1103/PhysRevX.8.021013
13. Rakovszky T, Pollmann F, von Keyserlingk CW (2017) Diffusive hydrodynamics of out-of-time-ordered correlators with charge conservation. arXiv:1710.09827
14. Zhou T, Nahum A (2018) Emergent statistical mechanics of entanglement in random unitary circuits. arXiv:1804.09737
15. Kitaev AY (2015) KITP talk: a simple model of quantum holography. http://online.kitp.ucsb.edu/online/entangled15/kitaev/
16. Maldacena J, Stanford D (2016) Remarks on the Sachdev-Ye-Kitaev model. Phys Rev D 94:106002. https://doi.org/10.1103/PhysRevD.94.106002
17. Aleiner IL, Faoro L, Ioffe LB (2016) Microscopic model of quantum butterfly effect: out-of-time-order correlators and traveling combustion waves. Ann Phys (NY) 375:378–406. https://doi.org/10.1016/j.aop.2016.09.006
18. Khemani V, Huse DA, Nahum A (2018) Velocity-dependent Lyapunov exponents in many-body quantum, semi-classical, and classical chaos. arXiv:1803.05902
19. Shenker SH, Stanford D (2014) Black holes and the butterfly effect. J High Energy Phys 2014:67. https://doi.org/10.1007/JHEP03(2014)067
20. Lieb EH, Robinson DW (1972) The finite group velocity of quantum spin systems. Commun Math Phys 28:251–257. https://doi.org/10.1007/BF01645779
21. Schneider U, Hackermuller L, Will S, Best T, Bloch I, Costi TA, Helmes RW, Rasch D, Rosch A (2008) Metallic and insulating phases of repulsively interacting fermions in a 3D Optical lattice. Science 322:1520–1525. https://doi.org/10.1126/science.1165449
22. Hackermuller L, Schneider U, Moreno-Cardoner M, Kitagawa T, Best T, Will S, Demler E, Altman E, Bloch I, Paredes B (2010) Anomalous expansion of attractively interacting fermionic atoms in an optical lattice. Science 327:1621–1624. https://doi.org/10.1126/science.1184565, arXiv:0912.3592

23. Braun S, Schneider U, Rasch D, Mandt S, Hackermüller L, Best T, Rosch A, Ronzheimer JP, Bloch I, Will S, Demler E (2012) Fermionic transport and out-of-equilibrium dynamics in a homogeneous Hubbard model with ultracold atoms. Nat Phys. https://doi.org/10.1038/nphys2205
24. Vardhan S, De Tomasi G, Heyl M, Heller EJ, Pollmann F (2017) Characterizing time irreversibility in disordered fermionic systems by the effect of local perturbations. Phys Rev Lett 119:016802. https://doi.org/10.1103/PhysRevLett.119.016802
25. Landau LD, Lifshitz EM (1977) Quantum mechanics: non-relativistic theory, 3rd edn. Pergamon Press
26. Serbyn M, Abanin DA (2017) Loschmidt echo in many-body localized phases. Phys Rev B 96:1–10. https://doi.org/10.1103/PhysRevB.96.014202, arXiv:1701.07772
27. Lee J, Kim D, Kim DH (2018) Typical growth behavior of the out-of-time-ordered commutator in many-body localized systems. arXiv:1812.00357
28. Hetterich D, Serbyn M, Domínguez F, Pollmann F, Trauzettel B (2017) Noninteracting central site model: localization and logarithmic entanglement growth. Phys Rev B 96:104203. https://doi.org/10.1103/PhysRevB.96.104203, arXiv:1701.02744v2
29. Yao NY, Grusdt F, Swingle B, Lukin MD, Stamper-Kurn DM, Moore JE, Demler EA (2016b) Interferometric approach to probing fast scrambling. arXiv:1607.01801
30. Garttner M, Bohnet JG, Safavi-Naini A, Wall ML, Bollinger JJ, Rey AM (2017) Measuring out-of-time-order correlations and multiple quantum spectra in a trapped-ion quantum magnet. Nat Phys 13:781–786. https://doi.org/10.1038/NPHYS4119

# Chapter 6
# Interactions

Our model can be modified in a variety of ways to include additional interactions that take it away from the free fermion limit discussed in previous chapters. Here we consider a subset of such extensions, focussing on the 1D case because of numerical limitations. Terms that can be added to the Hamiltonian (2.1) fall into two classes depending on whether they give dynamics to conserved charges. Away from the free fermion limit it is not possible to use determinant methods, and instead we have to resort to exact diagonalization and Krylov subspace methods to calculate the time evolution, see Appendix C.

## 6.1 Conserved Charges: Many-Body Localization

Let us first consider those additional terms in the Hamiltonian that commute with the charges $\hat{q}_j$. The conservation of these charges means that they still play the role of an effective binary potential for the fermions and we can still make a direct connection to localization physics. Such terms include fermion density-density interactions, and longitudinal field terms,

$$\Delta \sum_{\langle jk \rangle} \hat{n}_j \hat{n}_k, \quad \text{and} \quad B_x \sum_j \hat{\sigma}^x_{jk}, \tag{6.1}$$

respectively, where $\hat{n}_j = \hat{f}^\dagger_j \hat{f}_j$. The fact that these commute with the charges $\hat{q}_j = \hat{\sigma}^x_{j-1,j} \hat{\sigma}^x_{j,j+1} (-1)^{\hat{n}_j}$ follows from the commutation relations $[\hat{\sigma}^x_j, \hat{\sigma}^x_k] = 0$ and $[\hat{n}_j, \hat{n}_k] = 0$, for all $j$ and $k$.



A. Smith, *Disorder-Free Localization*, Springer Theses,
https://doi.org/10.1007/978-3-030-20851-6_6

### 6.1.1 XXZ Spin Chain

First, we consider adding nearest-neighbour density-density interactions between the fermions. Up to constant terms, the Hamiltonian can be written as

$$\hat{H} = -J \sum_{\langle jk \rangle} \hat{\sigma}^z_{jk} \hat{f}^\dagger_j \hat{f}_k - h \sum_j \hat{\sigma}^x_{j-1,j} \hat{\sigma}^x_{j,j+1} + \Delta \sum_j (2\hat{n}_j - 1)(2\hat{n}_{j+1} - 1). \quad (6.2)$$

Importantly, these density interactions also commute with the plaquette operators $\hat{B}_p$ which means that the duality mapping of the spins (2.5) is still valid. Under the transformation to $c$-fermions and charges, the first two terms transform as before and the fermion interactions have the same form since $\hat{f}^\dagger_j \hat{f}_j \equiv \hat{c}^\dagger_j \hat{c}_j$. The Hamiltonian in the dual language then reads

$$\hat{H} = -J \sum_j \left( \hat{c}^\dagger_j \hat{c}_{j+1} + \text{H.c} \right) + h \sum_j \hat{q}_j (2\hat{n}_j - 1) + \Delta \sum_j (2\hat{n}_j - 1)(2\hat{n}_{j+1} - 1). \quad (6.3)$$

Note that this Hamiltonian is of exactly the same form as the model for many-body localization presented in Eq. (1.24), with the exception that the potential is here determined by the conserved charges. We can then use a Jordan-Wigner transformation (see Sect. 1.3.2) to cast the Hamiltonian in the form

$$\hat{H}_{XXZ} = -J \sum_j (\hat{S}^+_j \hat{S}^-_{j+1} + \hat{S}^-_j \hat{S}^+_{j+1}) + h \sum_j \hat{q}_j \hat{S}^z_j + \Delta \sum_j \hat{S}^z_j \hat{S}^z_{j+1}, \quad (6.4)$$

which is an XXZ Hamiltonian describing a spin chain with a binary potential set by the charge configuration $\{q_j\} = \pm 1$. This XXZ Hamiltonian with quenched disorder serves as one of the paradigmatic models of MBL. Although in the context of MBL this model it is usually studied with uniformly sampled disorder [1, 2], it has also been studied in the case of binary disorder [3, 4]. Here we observe the behaviour usually found in MBL phases despite the Hamiltonian (6.2) being disorder-free, as are the initial states that we consider.

Let us consider the charge density wave initial state, which we considered in Sect. 3.1, see Fig. 6.1a. We find that, as in the non-interacting case, the density imbalance $\Delta\rho(t)$ saturates at a non-zero value indicating the persistent memory of the initial state due to localization. For small interactions, $\Delta = 0.1J$, the asymptotic value is close to the value found in the non-interacting case. As the interaction strength is increased it also acts to stabilise the charge density wave and leads to an increase of the asymptotic value, as can already be seen for $\Delta = 0.3J$. We also observe that the interactions have the effect of damping the fluctuations around this asymptotic value, which is evident over the time scales shown in Fig. 6.1a.

Next we can consider the von Neumann entanglement entropy, shown in Fig. 6.1b. Here we see a qualitative change (compared to the non-interacting case) in the entanglement entropy growth following the initial area-law plateau. While in the

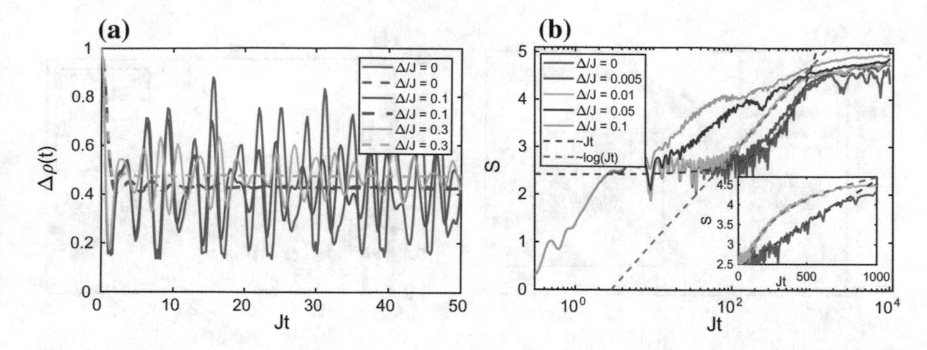


**Fig. 6.1** Quantum quench from an initial charge density wave state for $h/J = 20$. **a** Density imbalance $\Delta\rho(t) \propto \sum_j |\langle 0|\hat{n}_j(t) - \hat{n}_{j+1}(t)|0\rangle|$ and the time-averaged value of $\frac{1}{t}\int_0^t \mathrm{d}\tau\, \Delta\rho(\tau)$ (dashed lines) after the same quench. **b** Von Neumann entanglement entropy computed using ED for $N = 12$ sites (thin, light) for various values of $\Delta$ shown on a semi-log scale. The spatial bipartition is taken along the central bond. (inset) The same data on a linear scale for $\Delta/J = 0, 0.01$. Dashed lines show fitted linear and logarithmic curves. The figure is reproduced from the journal Physical Review B [5]

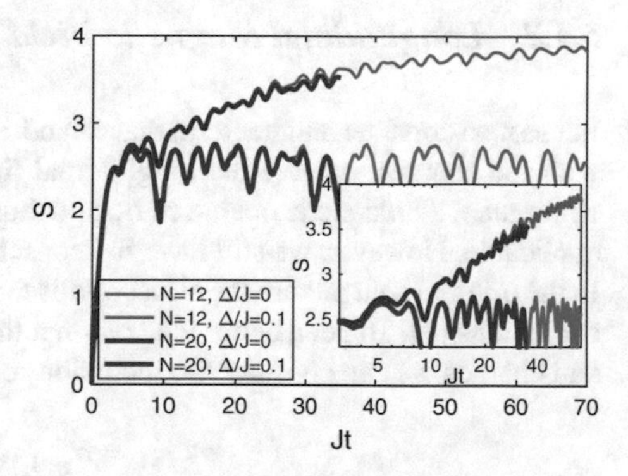


**Fig. 6.2** Comparison of entanglement using ED with that obtained using matrix product state methods, with additional density interactions $\Delta \sum_{\langle jk\rangle} \hat{n}_j \hat{n}_k$. The ED results are for a chain with $N = 12$ and the matrix product state results for $N = 20$. Inset shows the same data on a logarithmic scale. The figure is reproduced from the journal Physical Review Letters [6]

non-interacting case we have linear growth followed by saturation, in the presence of density-density interactions we observe earlier but slower logarithmic growth, as shown by the dashed lines in Fig. 6.1b, and in the inset. This logarithmic behaviour, which sets in at times $\sim\Delta/J$, is consistent with the phenomenology of MBL, whereas the linear growth begins at times $\sim(h/J)^2$. We also used a matrix product state numerical method (using iTensor [7]) to compute the entanglement for a system of $N = 20$ sites, see Fig. 6.2. Due to the large bond dimension needed, we could not go to long times but we observed the logarithmic growth over the time scales that we could access and found agreement with the ED results. This confirms that during this growth the entanglement entropy is independent of system size.

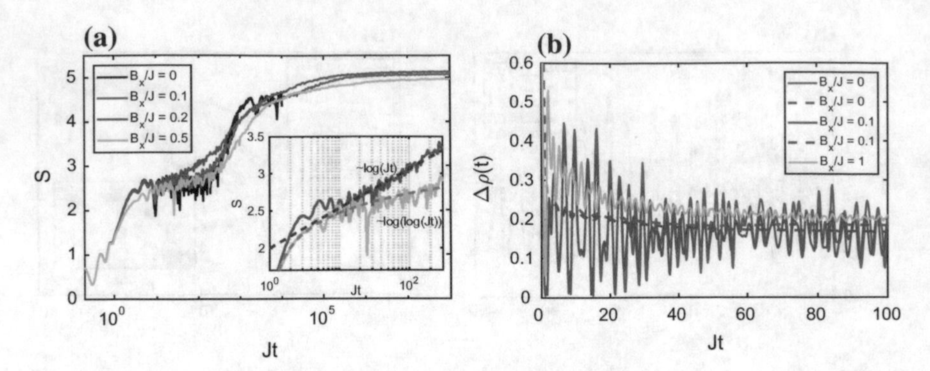


**Fig. 6.3** **a** Entanglement entropy after a quantum quench for various values of the longitudinal field $B_x$ shown on a semi-log scale. A window of the same data for $B_x/J = 0.2, 0.5$ is given in the inset. Dashed lines correspond to $\log(t)$ and $\log(\log(t))$ behaviour. **b** Density imbalance after a quench from a charge density wave. Results obtained using ED for $N = 12$ sites. The figure is reproduced from the journal Physical Review B [5]

### *6.1.2 Longitudinal Magnetic Field*

Let us next consider another term that can be added to the Hamiltonian that commutes with the charges, namely the longitudinal field. Unfortunately, this term does not commute with plaquette operators $\hat{B}_p$ and thus the duality mapping (2.5) is no longer applicable. However, we still have the conserved charges $\hat{q}_j = \hat{\sigma}^x_{j-1,j}\hat{\sigma}^x_{j,j+1}(-1)^{\hat{n}_j}$. In the original spin picture the effect of this term in the Hamiltonian is to confine spin excitations [8]. To get a better idea of what this term looks like in terms of fermions and charges we can consider the recursion relation

$$\hat{\sigma}^x_{j,j+1} = \hat{\sigma}^x_{j-1,j}\hat{q}_j(-1)^{\hat{n}_j}, \tag{6.5}$$

which follows from the form of the conserved charges. For a 1D chain with open boundary conditions we can then write

$$\sum_j \hat{\sigma}^x_{j,j+1} = \hat{\sigma}^x_{1,2} \sum_j \prod_{i<j} \hat{q}_{i+1}(1 - 2\hat{n}_{i+1}), \tag{6.6}$$

that is, the longitudinal field results in effective long range density interactions for the fermions. Note that the presence of these long range interactions is related to the fact that $\sum_{\langle jk \rangle} \hat{\sigma}^x_{jk}$ is analogous to $\mathbf{E}^2$ in QED, which in the discrete theory leads to the confinement of charges. Whereas, the Ising coupling $\sum_j \hat{\sigma}^x_{j-1,j}\hat{\sigma}^x_{j,j+1}$ corresponds to $\nabla \cdot \mathbf{E}$ and the charges are deconfined.

In Fig. 6.3a we present the results for the entanglement entropy with the Hamiltonian having a longitudinal field term whose strength is controlled by $B_x$. The results show a rich behaviour. In particular one can notice two new qualitative features.

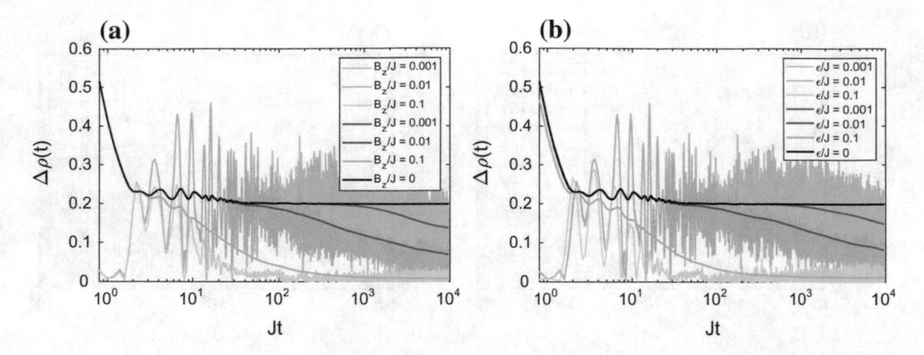


**Fig. 6.4** Density imbalance after a quantum quench from a CDW with dynamical charges. We study a system with $N = 10$ sites and periodic boundary conditions for $h = J$. The light curves correspond to the imbalance $\Delta\rho(t)$ and the dark thick lines are the time averaged value $\frac{1}{t}\int_0^t d\tau\, \Delta\rho(\tau)$. **a** With an additional transverse field $B_z \sum_i \hat{\sigma}^z_{i,i+1}$. **b** Additional fermion hopping $\epsilon \sum_{\langle ij\rangle} \hat{f}_i^\dagger \hat{f}_j$. Dynamics is computed using Krylov subspace methods, see Appendix C. The figure is reproduced from the journal Physical Review B [5]

For small $B_x/J = 0.2$, we observe logarithmic entanglement growth at a time scale set by $\sim B_x/J$ similar to the MBL behaviour observed above. However, for larger $B_x/J = 0.5$ we find a slower growth, which can be fitted by $\log(\log(t))$ as shown in the inset. Similar sub-logarithmic growth was observed in Ref. [9] where they consider the disorder-free mechanism for localization in the $U(1)$ lattice Schwinger model. Furthermore, for small $h$ the interactions generate additional entanglement compared to the non-interacting results, whereas for larger $B_x$ the entanglement is reduced. Looking at the density imbalance $\Delta\rho(t)$ after a quench from a charge density wave, shown in Fig. 6.3b, we again see that interactions have a damping effect on the fluctuations and that a strong enough field stabilises the charge density wave.

## 6.2 Dynamical Charges: Quasi-MBL

In the second category, i.e., those terms that do not commute with charges and generate their dynamics, we consider three types of terms

$$B_z \sum_j \hat{\sigma}^z_{j,j+1}, \qquad \epsilon \sum_{\langle ij\rangle} \hat{f}_i^\dagger \hat{f}_j, \qquad h_z \sum_j \hat{\sigma}^z_{j-1,j}\hat{\sigma}^z_{j,j+1}. \tag{6.7}$$

The localization behaviour studied in previous chapters relied on the presence of static charges $\hat{q}_j$, which act as an effective disorder potential. It is therefore a natural question to ask what happens when these charges have dynamics.

Figure 6.4 shows the effect of these additional interactions on the density imbalance after a quench from the charge density wave initial state. Figure 6.4a–b clearly

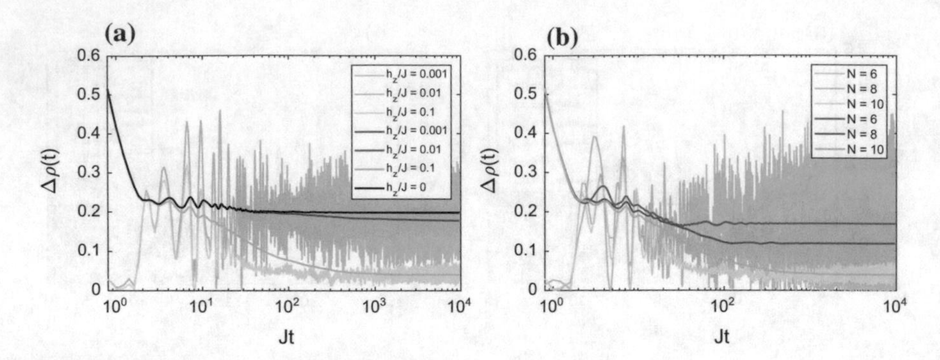


**Fig. 6.5** Finite size scaling of the asymptotic density imbalance in the presence of additional Ising coupling $h_z \sum_i \hat{\sigma}^z_{j-1,j}\hat{\sigma}^z_{j,j+1}$. We use system sizes $N = 6, 8, 10$ with $h_z = 0.1J$ and $h = J$. Light curves correspond to the density imbalance $\Delta\rho(t)$ and the dark and thick curves show the time averaged value of $\frac{1}{t}\int_0^t d\tau\,\Delta\rho(\tau)$. See Appendix C for details of the Krylov subspace numerical method used. The figure is reproduced from the journal Physical Review B [5]

show that the introduction of $B_z$ and $\epsilon$ leads to the decay of this imbalance, and ultimately $\Delta\rho(t)$ vanishes. One can also see that the time scale at which the results significantly deviate from the $B_z = \epsilon = 0$ case is determined by $B_z^{-1}$ and $\epsilon^{-1}$ respectively. A qualitative difference between these two terms is that the fermion hopping $\epsilon$ to lowest order modifies $J$. This effect can be seen as an increase in the frequency of oscillations at $\epsilon = 0.1$. Beyond this point, there are only quantitative differences between the two cases, and the time averaged values look similar to the eye.

A different phenomenology is observed in the case of the $z$-Ising coupling $h_z$, shown in Fig. 6.5a. In this case there is little appreciable deviation in the time averaged value for $h_z/J = 0.001, 0.01$, other than the damping of the oscillations. When the coupling is increased to $h_z = 0.1J$ we finally see decay of the imbalance, but it does not convincingly vanish. This behaviour can be understood by considering the (anti-)commutation relations of $\hat{\sigma}^z_{j,j+1}$ with the charges $\hat{q}_j$, which are

$$\begin{aligned} \{\hat{\sigma}^z_{j,j+1}, \hat{q}_k\} &= 0, \quad k = j, j+1, \\ [\hat{\sigma}^z_{j,j+1}, \hat{q}_k] &= 0, \quad k \neq j, j+1. \end{aligned} \tag{6.8}$$

We can then make an identification with spin operators $\hat{q}_j \to \hat{q}^z_k$ and $\hat{\sigma}^z_{j,j+1} \to \hat{q}^x_j \hat{q}^x_{j+1}$, that is, the transverse field induces an effective Ising coupling between the charges. For the transverse field we are then precisely in the framework of heavy-light mixtures which are generally believed to become ergodic at long-times. However, the Ising coupling $\hat{\sigma}^z_{j-1,j}\hat{\sigma}^z_{j,j+1}$ maps to the next-nearest neighbour Ising coupling $\hat{q}^x_{j-1}\hat{q}^x_{j+1}$ for the charges. Since the lattice is bipartite, the charges interact separately on two disconnected sublattices. Importantly, this means that the charges—and by extension the effective disorder potential—on neighbouring sites do not become correlated through direct interaction but only through higher order processes. This leads to increased persistence of localization seen in Fig. 6.5a.

Since we still have a heavy-light mixture with the addition of the $z$-Ising coupling, we can ask whether this additional persistence survives in the thermodynamic limit at long times, which would be in contrast with the standard phenomenology of these systems. Figure 6.5b shows the density imbalance as a function of the system size, which seems to suggest that we also lose localization in this case in the thermodynamic limit, consistent with Ref. [10].

## References


1. Žnidarič M, Prosen T, Prelovšek P (2008) Many-body localization in the Heisenberg XXZ magnet in a random field. Phys Rev B 77:064426. https://doi.org/10.1103/PhysRevB.77.064426
2. Bardarson JH, Pollmann F, Moore JE (2012) Unbounded growth of entanglement in models of many-body localization. Phys Rev Lett 109:017202. https://doi.org/10.1103/PhysRevLett.109.017202
3. Andraschko F, Enss T, Sirker J (2014) Purification and many-body localization in cold atomic gases. Phys Rev Lett 113:217201. https://doi.org/10.1103/PhysRevLett.113.217201
4. Tang B, Iyer D, Rigol M (2015) Quantum quenches and many-body localization in the thermodynamic limit. Phys Rev B 91. https://doi.org/10.1103/PhysRevB.91.161109
5. Smith A, Knolle J, Moessner R, Kovrizhin DL (2018a) Dynamical localization in $Z_2$ lattice gauge theories. Phys Rev B 97:245137. https://doi.org/10.1103/PhysRevB.97.245137
6. Smith A, Knolle J, Moessner R, Kovrizhin DL (2017b) Absence of ergodicity without quenched disorder: from quantum disentangled liquids to many-body localization. Phys Rev Lett 119:176601. https://doi.org/10.1103/PhysRevLett.119.176601
7. Stoudenmire EM, White SR, Others, iTensor. http://itensor.org
8. Kormos M, Collura M, Takács G, Calabrese P (2016) Real-time confinement following a quantum quench to a non-integrable model. Nat Phys 13:246–249. https://doi.org/10.1038/nphys3934
9. Brenes M, Dalmonte M, Heyl M, Scardicchio A (2018) Many-body localization dynamics from gauge invariance. Phys Rev Lett 120:030601. https://doi.org/10.1103/PhysRevLett.120.030601
10. Papić Z, Stoudenmire EM, Abanin DA (2015) Many-body localization in disorder-free systems: the importance of finite-size constraints. Ann Phys (N Y) 362, 714–725. https://doi.org/10.1016/j.aop.2015.08.024

# Chapter 7
# Experimental Proposal

Quantum simulators are well-controlled quantum systems used for simulating theoretical quantum models and hold the promise of studying physics beyond what is accessible by classical computation. Due to the remarkable recent experimental advances in the control of isolated quantum systems, they are now becoming a reality, and provide a powerful setting for studying strongly coupled quantum systems. These advances have come from a wide range of settings including superconducting chips [1], photonic quantum circuits [2], and notably trapped ions [3, 4]. In this Chapter we focus on optical lattices in cold atom experiments like those recently used to study many-body localization phenomena in a large two-dimensional lattice [5].

Until now, the quantum simulation of LGT has been restricted to one dimensional systems because of the strong demand on control (fidelity) and sheer number of qubits required for these simulations. Here we show a minimal setting for simulating the dynamics of a LGT in two dimensions. This model provides an ideal candidate for experimental implementation for three main reasons. Firstly, it can be reduced to free fermions and thus can be benchmarked against classical simulations. Via duality transformations we show that dynamical correlation functions of the gauge fields can be directly mapped to local impurity quenches of free fermionic systems. Secondly, even in the free fermion limit, it has been shown to display novel phenomenology of disorder-free localization [6–8], and can easily be perturbed away from this 'integrable' limit where classical computation is no longer applicable. And thirdly, the measurement of correlators can be implemented with current technology in cold atomic gases [5] and we provide simple protocols based on Ramsey interferometry [9–12].


A. Smith, *Disorder-Free Localization*, Springer Theses,
https://doi.org/10.1007/978-3-030-20851-6_7

## 7.1 A Minimal $\mathbb{Z}_2$ Lattice Gauge Theory with Fermionic Matter

A long time goal would be the efficient quantum simulation of interacting quantum field theories, for example the paradigmatic Hamiltonian of Quantum Electrodynamics (QED) with minimally coupled fermion fields. Progress towards this goal is already being made, for instance in the digital simulation of the lattice Schwinger model using 10 trapped ions [4]. Here we present our model introduced in Eq. (2.1) as an example of a minimal $\mathbb{Z}_2$ lattice gauge theory with fermionic matter. In the setting of cold atom experiments we propose a protocol for measuring correlations of the gauge field in two-dimensions. The setup is similar to those studied in the context of MBL [5] and is possible to implement with current technology.

Let us re-introduce our model from a different perspective. We will start with a slightly modified version of the $U(1)$ LGT from Eq. (1.48)

$$\hat{H} = -J\sum_{\langle xy\rangle} s_{xy}(\hat{\psi}_x^\dagger \hat{U}_{xy}\hat{\psi}_y + \text{H.c.}) + m\sum_x s_x\hat{\psi}_x^\dagger\hat{\psi}_x - h\sum_x \cos([\nabla\cdot\hat{E}]_x) - K\sum_p(\hat{U}_p + \hat{U}_p^\dagger). \tag{7.1}$$

Note the non-standard term $\cos([\nabla\cdot\hat{E}]_x)$, which corresponds to $(\nabla\cdot\hat{E})^2$ in the continuum. The discrete divergence for square or cubic lattices is defined to be $[\nabla\cdot\hat{E}]_x = \sum_\mu(\hat{E}_{x,x+\mu} - \hat{E}_{x,x-\mu})$, where $\mu$ are lattice vectors. We will then consider only spinless fermions and so we can drop the sign factors $s_{xy}$ and $s_x$, which encode the structure of spinor space. We also replace the $U(1)$ gauge degrees of freedom with $\mathbb{Z}_2$, that is, the gauge fields have eigenvalues $\hat{U}_{xy} = \pm 1$ and $\hat{E}_{x,y} = 0, 1$. Let us then slightly change notation with $\hat{\psi}_x \to \hat{f}_j$ for the spinless fermion operators, and $\hat{U}_{xy} \to \hat{\sigma}_{jk}^z$ and $e^{i\pi\hat{E}_{x,y}} \to \hat{\sigma}_{jk}^x$ for the gauge fields, where now $j, k$ refer to physical sites. The Hamiltonian then becomes

$$\hat{H} = -J\sum_{\langle ij\rangle}\hat{\sigma}_{ij}^z\hat{f}_i^\dagger\hat{f}_j + m\sum_j\hat{f}_j^\dagger\hat{f}_j - h\sum_i\hat{A}_i - K\sum_p\hat{B}_p, \tag{7.2}$$

with the star and plaquette operators

$$\hat{A}_i = \prod_{j\in +_i}\hat{\sigma}_{ij}^x, \qquad \hat{B}_p = \prod_{\text{plaquette } p}\hat{\sigma}_{jk}^z. \tag{7.3}$$

where the gauge field operators $\hat{\sigma}_{jk}^z$ and $\hat{\sigma}_{jk}^x$ are the Pauli matrices and the star operator corresponds to $\cos([\nabla\cdot\hat{E}]_j)$. The conserved quantities are $\hat{q}_j = \hat{A}_j(-1)^{\hat{n}_j}$ as before, which is a generalized Gauss' law, relating the divergence of the gauge field to the fermion density. See Refs. [13–16] for more details about this transition from continuum $U(1)$ to discrete $\mathbb{Z}_2$. In the following we will only consider quenches from an initial state $|\Psi\rangle$ that is invariant under the action of plaquette operators, $\hat{B}_p|\Psi\rangle = |\Psi\rangle$, and has fixed fermion filling. Since the plaquette operators are conserved, the plaquette term is a constant under the Hamiltonian evolution, as is the mass term due to the fixed fermion number, and so we drop these terms from the Hamiltonian and return to the model in Eq. (2.1). As before, the important step in our analysis is a duality transformation of the spins, which is valid since we have fixed $\hat{B}_p = 1$.

In this chapter we will concentrate on the dynamics of the gauge field in two dimensions. The quantities that we will consider are the time-dependent correlation functions of the gauge sector

$$\langle \hat{\sigma}^z_{jk}(t)\hat{\sigma}^z_{lm}(t)\rangle_c = \langle \hat{\sigma}^z_{jk}(t)\hat{\sigma}^z_{lm}(t)\rangle - \langle \hat{\sigma}^z_{jk}(t)\rangle\langle \hat{\sigma}^z_{lm}(t)\rangle, \tag{7.4}$$

after a quantum quench. In the following, we propose a protocol for their quantum simulation in cold atom experiments and present some numerical results.

## 7.2 The Quench Protocol

We consider a quench protocol where the spins and fermions are prepared in a particular initial state with respect to which we wish to calculate the dynamics of correlations of the form in Eq. (7.4). We prepare our spins and fermions in an initial state that is a tensor product of the $z$-polarized spin state and a fermion Slater determinant at half-filling $|\Psi\rangle = |\uparrow\uparrow\cdots\rangle \otimes |\psi\rangle$. The Slater determinant corresponds to fermions in a Fermi-sea configuration for the Hamiltonian $\hat{H}_{\text{FS}} = -\sum_{\langle ij\rangle} \hat{f}^\dagger_i \hat{f}_j$. Importantly, these initial states take the form

$$|\Psi\rangle = \frac{1}{\sqrt{2^{N-1}}} \sum_{\{q_i\}=\pm 1}{}' |q_1 q_2 \cdots q_N\rangle \otimes |\psi\rangle, \tag{7.5}$$

in terms of the charges $\hat{q}$ and the fermions $\hat{c}$, where we note that $|\psi\rangle_f = |\psi\rangle_c$, see Sect. 2.1 for more details.

Let us now consider the calculation of the correlator. One of the simplest components of the connected spin correlator that we wish to calculate is the average local magnetisation

$$\langle \hat{\sigma}^z_{jk}(t)\rangle = \langle \Psi| e^{i\hat{H}t} \hat{\tau}^x_j \hat{\tau}^x_k e^{-i\hat{H}t} |\Psi\rangle. \tag{7.6}$$

We can then work with the Hamiltonian for a fixed configuration of charges $\{q_i\} = \pm 1$, defined in Eq. (2.22), which we write down again for convenience

$$\hat{H}(q) = -J \sum_{\langle jk\rangle} \hat{c}^\dagger_j \hat{c}_k + 2h \sum_j q_j (\hat{c}^\dagger_j \hat{c}_j - 1/2). \tag{7.7}$$

In terms of these single particle Hamiltonians we can write the expectation value as

$$\langle \hat{\sigma}^z_{jk}(t)\rangle = \frac{1}{2^{N-1}} \sum_{\{q_i\}=\pm 1}{}' \langle \psi| e^{i\hat{H}(q)t} e^{-i\hat{H}_{jk}(q)t} |\psi\rangle, \tag{7.8}$$

as we also found earlier in Sect. 2.2. This is in the form of a disorder-averaged Loschmidt echo, where the charges at site $j, k$ are flipped between the forward and backward evolution. Repeating the same arguments we find the expression

$$\langle \hat{\sigma}^z_{jk}(t)\hat{\sigma}^z_{lm}(t)\rangle = \frac{1}{2^{N-1}} \sum_{\{q_i\}=\pm 1}' \langle \psi | e^{i\hat{H}(q)t} e^{-i\hat{H}_{jklm}(q)t} |\psi\rangle, \tag{7.9}$$

for the two-point correlator, where we flip the four charges at sites $j, k, l, m$ between forward and backward evolution. The dynamical correlation function of the gauge field directly corresponds to a local quantum quench of a (free) fermionic Hamiltonian. In the following section we discuss how the latter can be efficiently simulated in a system of cold atoms.

Numerically, the free fermion Loschmidt echo appearing in these expressions can be efficiently computed using determinants, see Appendix B. For our particular setup the determinants in Eq. (7.8) take the form

$$\langle \psi | e^{i\hat{H}(q)t} e^{-i\hat{H}_{jk}(q)t} |\psi\rangle = \det[V^\dagger U^\dagger(q) U_{jk}(q) V], \tag{7.10}$$

where $U(q) = e^{-iH(q)t}$ is the matrix exponential of the single-particle Hamiltonian matrix $H(q)$, and similarly for $U_{jk}(q)$ and $H_{jk}(q)$. $V$ is a rectangular matrix that has as its columns the $N/2$ filled single particle eigenvectors of the Hamiltonian $\hat{H}_{\mathrm{FS}} = -\sum_{\langle ij\rangle} \hat{f}_i^\dagger \hat{f}_j$. The determinants for the two-point correlators (7.9) take a similar form but with $H_{jklm}(q)$ replacing $H_{jk}(q)$.

## 7.3 Quantum Simulation and Experimental Setup

The experimental setup that we have in mind is shown schematically in Fig. 7.1. We consider a two dimensional square optical lattice half-filled with fermions, which have a nearest neighbour hopping amplitude between the sites of this lattice. The fermions are then subjected to a disordered binary potential—similar to the quantum gas microscope set up used in Ref. [5]—and we have two or four impurity spins that control the potential flips for the Loschmidt echo we want to calculate [10–12]. These impurities can be trapped in a potential with depth essentially independent of the one felt by the fermions. It is possible for the fermions to interact strongly with one of the two spin states $|\uparrow_j\rangle$ and weakly with the other $|\downarrow_j\rangle$, effectively turning on/off of the potential on that site.

As explained above, the correlators we wish to calculate correspond to Loschmidt echoes. We implement these using Ramsey interferometry, which we now briefly outline. When the impurities are in the up state they interact with the host fermions, which see a local potential; in the down state they are effectively decoupled. Further details on experimental implementations can be found in [9–12]. Now we introduce a composite two state system, called the control spin. For the average local magnetisation the impurity control spin controls two neighbouring charges, i.e. $|\Downarrow\rangle \leftrightarrow |\downarrow_j\downarrow_k\rangle$, $|\Uparrow\rangle \leftrightarrow |\uparrow_j\uparrow_k\rangle$, and for the two point correlator we have two pairs of impurity spins (as shown in Fig. 7.1), i.e., $|\Downarrow\rangle \leftrightarrow |\downarrow_j\downarrow_k\downarrow_l\downarrow_m\rangle, |\Uparrow\rangle \leftrightarrow |\uparrow_j\uparrow_k\uparrow_l\uparrow_m\rangle$. Using the average local magnetisation as an explicit example, the procedure is given by the steps in Box I.

**Initialise**—We initialise the state of the system so that the fermions are in the half-filled ground state $|\psi\rangle$ of $\hat{H}_{\text{FS}} = -\sum_{\langle ij\rangle} \hat{f}_i^\dagger \hat{f}_j$ and the control spin is in the state $|\Downarrow\rangle$.

$\pi/2$ **pulse**—At $t = 0$ we perform a $\pi/2$ pulse on the control spin such that the state of the system becomes $|\Psi\rangle = \frac{|\Uparrow\rangle+|\Downarrow\rangle}{\sqrt{2}}|\psi\rangle$.

**Evolve**—We let the system evolve, so that the state of the system at time $t$ is given by

$$|\Psi(t)\rangle = \frac{1}{\sqrt{2}}\left(e^{-i\hat{H}(q)t}|\Downarrow\rangle|\psi\rangle + e^{-i\hat{H}_{jk}(q)t}|\Uparrow\rangle|\psi\rangle\right). \quad (7.11)$$

$\pi/2$ **pulse**—At time $t$ we perform the reverse $\pi/2$ pulse on the control spin so that we can measure in the natural basis of impurity spins.

**Measure**—We measure $\hat{S}^z$ of the control spin. This measurement, once averaged over experimental realisations gives

$$\langle\hat{S}^z\rangle = \text{Re}\langle\psi|e^{i\hat{H}(q)t}e^{-i\hat{H}_{jk}(q)t}|\psi\rangle. \quad (7.12)$$

**Box I.** Outline of the experimental procedure for simulating the spin correlators in Eq. (7.4)

This full procedure must then be performed for and averaged over different disorder realisations. For the two-point correlator, the control spin corresponds to two pairs of impurities, which amounts to replacing $\hat{H}_{jk}(q)$ by $\hat{H}_{jklm}(q)$. These correlators are self-averaging and therefore it is sufficient to average over a small random subset of disorder realisations. Because the spin correlators must be real, this procedure amounts precisely to the calculations we wish to perform in Eq. (7.4).

## 7.4 Numerical Results

We now show some numerical results for a $15 \times 14$ square lattice for $h/J = 0.7, 2$ where we have averaged over 1000 random charge configurations. In Fig. 7.2 we show the time dependence of the connected two-point spin correlator (7.4) as a function of separation between the two spins along the horizontal and diagonal cuts indicated by dashed lines in Fig. 7.3. Two main features common to all four plots are the linear light-cone spreading and the eventual decay of all spatial correlations.

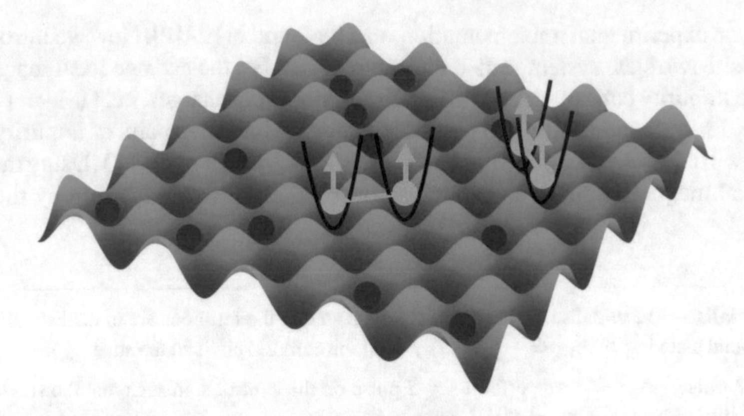

**Fig. 7.1** Schematic of the experimental protocol. Fermions (purple balls) are trapped in an optical lattice (blue surface) within which they are able to hop with Hamiltonian Eq. (7.7) with a predetermined binary potential set by charges $\{q_i\} = \pm 1$, except at particular sites where we have impurity spins (yellow). These impurity spins are localized in separately controlled, much deeper wells and play the role of the potential on that site, with spin up being positive and spin down being negative. To calculate physical spin-spin correlators we must control four impurity spins, which are paired along the bond associated with the physical spin. The correlators are then calculated using the Loschmidt echo protocol defined in the main text, which involves a $\pi/2$-rotation and measurement of these spins. The figure is reproduced from the journal Quantum Science and Technology [17]

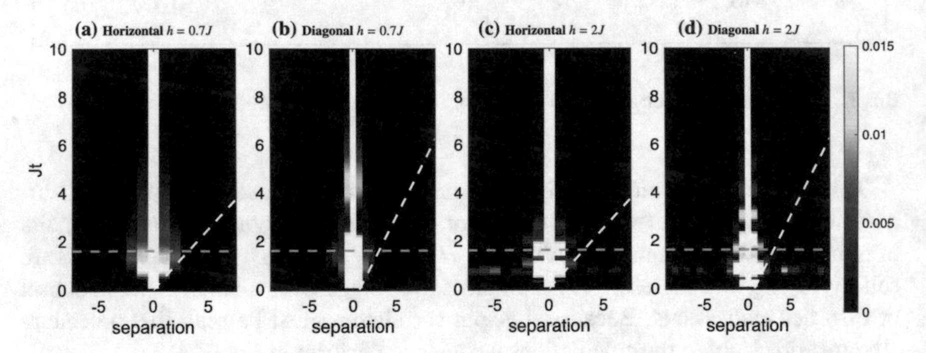


**Fig. 7.2** Absolute value of the connected spin-spin correlator $|\langle \hat{\sigma}_b^z(t)\hat{\sigma}_0^z(t)\rangle_c|$ for different cuts across the system as a function of time and separation between the spin $\hat{\sigma}_b^z$ and the central bond spin $\hat{\sigma}_0^z$. Results are shown for two different values of $h/J = 0.7, 2$. The horizontal and diagonal cuts are those shown in Fig. 7.3. The blue dashed line indicates the time slice that is shown in Fig. 7.3. The white dashed line indicates the propagation of the light-cone with velocity $v = 2J$. The figure is reproduced from the journal Quantum Science and Technology [17]

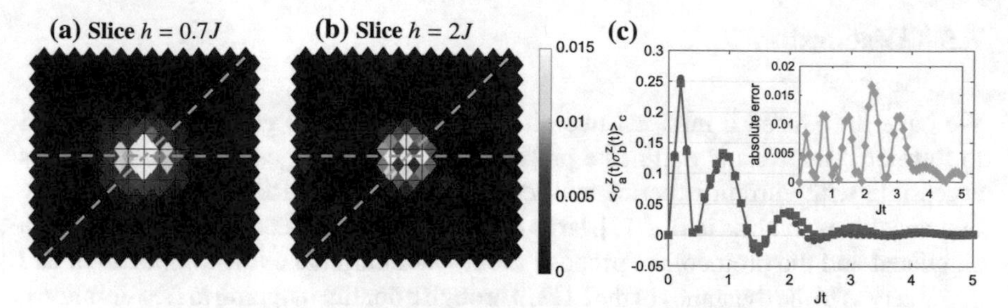


**Fig. 7.3** **a–b** Spatially resolved absolute value of the connected spin-spin correlator $|\langle\hat{\sigma}_b^z(t)\hat{\sigma}_0^z(t)\rangle_c|$. The spin $\hat{\sigma}_0^z$ is on the central bond and $\hat{\sigma}_b^z$ is taken on the other bonds of the $15 \times 14$ lattice for $Jt = 1.7$ and $h/J = 0.7, 2$. Superimposed in black is the lattice on which the fermions sit. The dynamics along the horizontal and diagonal cuts indicated by dashed lines are shown in Fig. 7.2. **c** Comparison of a nearest neighbour correlator $\langle\hat{\sigma}_b^z(t)\hat{\sigma}_0^z(t)\rangle_c$ along the diagonal indicated in Fig. 7.3a, b with different numbers of sampled disorder configurations for $h = 2J$. The blue curve corresponds to averaging over 1000 random realisations, while the red corresponds to only 50. (Inset) The absolute discrepency of 50 disorder realisations relative to 1000. The figure is reproduced from the journal Quantum Science and Technology [17]

The spreading of correlations is linear in all cases and has velocity $v = 2J$, which is the maximal group velocity of the fermions. This light-cone regime is short-lived due to the overall decay of spatial correlations. A notable difference between the horizontal and diagonal cuts is that the correlations between the neighbouring spins along the diagonal grows immediately, leading to a slightly offset light-cone. This is because the spins belong to the same star operators (see Fig. 7.3) and thus the correlations start growing from $t = 0$ with a rate set by $h$, as shown in Fig. 7.3c. Different behaviour appears when we increase $h$, where we see that the peaks in the correlations get much sharper along this light-cone and then are followed by decaying oscillations. The spatial pattern of correlations also changes as shown for a time slice $Jt = 1.7$ in Fig. 7.3. The extent of the spreading is greater for lower $h$. Furthermore, for lower $h$ we can clearly see the asymmetry due to the fact that the central bond is vertical and we don't have 90° rotational symmetry, whereas for $h = 2J$ this asymmetry seems to be smaller.

While for an exact simulation of the gauge field we would need to average over all possible configurations of the potential due to the charges, we require only a tiny fraction of this total number to obtain accurate results. For $h = 0.7J$, we see that the results have the correct symmetry and there is very little random noise. On the other hand, for $h = 2J$ we see more non-physical correlations, most notably as a stripe in Fig. 7.2c, d at around $Jt = 1$, which partly obscures the linear light-cone. We can also see a faint non-uniform random background in Fig. 7.3b. To remove these features we need to use a larger number of configurations. Despite this, qualitative results can be obtained with as few as 50 disorder configurations, as shown in Fig. 7.3c, for the nearest neighbour correlation along the diagonal with $h = 2J$. With so few samples we are still able to extract qualitative properties such as the immediate and sharp growth of correlations and the subsequent decaying oscillations.

## 7.5 Discussion

We have introduced a minimal two-dimensional $\mathbb{Z}_2$ lattice gauge theory coupled to fermionic matter and outlined a protocol for measuring gauge field correlations accessible with current experimental technology. In cold atom experiments using optical lattices, such as in Ref. [5], large two-dimensional systems have already been simulated and the protocol we propose should add minimal extra complication and allow access to the dynamics of the LGT. Through a duality mapping to free-fermions, we are able to perform efficient numerical computations, which allows benchmarking of the experiment. In the results that we have presented in this Chapter we can see some clear dynamical features, such as the light-cone spreading of correlations and their long-time decay.

There may be several practical issues regarding the difference between the experimental setup and the ideal model that we present but many of these can easily be accounted for. These include an effecitve smoothing of the binary potential, the shape of an additional trapping potential of the fermions, or the incomplete decoupling of the impurity spins. All of these can be included in the classical simulation and so it can be checked whether they introduce any additional physics. Furthermore, it is possible numerically to do scaling in both system size and the number of disorder realisations to pin down the accuracy that can be expected in such simulations. The results presented here already show that we can expect qualitative agreement with only a small number of samples.

While the dynamics of the model in Eq. (2.1) can be efficiently computed classically, there are several generalisations that render it truly interacting. In particular, there are terms that can be added to the Hamiltonian that commute with the charges, which means that the above mapping and experimental protocol are still valid. These were discussed in Sect. 2.3 and in Chap. 6. These included the nearest-neighbour density interaction $\sum_{\langle jk \rangle} \hat{n}_j \hat{n}_k$, which has the same form before and after the duality mapping. With this term, our model maps to an XXZ spin chain with a binary disorder potential. Another generalisation that can be considered is to add a spin degree freedom to the fermions, which is related to a slave-spin description of the Hubbard model [18, 19]. With the addition of interactions, classical computation can only access system sizes that fall far short of the thermodynamic limit in two dimensions. In experiment, however, these generalisations should not pose significant extra difficultly. Our model therefore provides an ideal setting for simulating LGT dynamics in two-dimensions beyond classical capabilities, with the bonus feature of a well defined free-fermion limit that can be reliably benchmarked.

### 7.5.1 Simulating Out-of-Time-Ordered Correlators

It may also be possible to extend the experimental protocol that we have described for simulating gauge field correlators to OTOCS, as discussed in Chap. 5. The main measurement technique would similarly be based on Ramsey interferometry but there would be additional technical difficulties associated with extra evolution backwards and forwards in time. We will briefly consider this possibility in a bit more detail here.

Let us consider the OTOC $C_{zz}(t)$ for the $z$-component of spin defined as $C_{zz}(t) = 1 - \text{Re}\, F_{zz}(t)$, with

$$F_{zz}(t) = \langle\Psi|\hat{\sigma}^z_{jk}(t)\hat{\sigma}^z_{lm}\hat{\sigma}^z_{jk}(t)\hat{\sigma}^z_{lm}|\Psi\rangle. \tag{7.13}$$

Using the mapping to free-fermions, in Sect. 5.1.1 we re-expressed this correlator in terms of a disorder-averaged double Loschmidt echo, namely

$$F_{zz}(t) = \frac{1}{2^{N-1}} \sum_{\{q_i\}=\pm 1}' \langle\psi|e^{i\hat{H}(q)t}e^{-i\hat{H}_{jk}(q)t}e^{i\hat{H}_{jklm}(q)t}e^{-i\hat{H}_{lm}(q)t}|\psi\rangle, \tag{7.14}$$

where $\hat{H}(q)$ is the free fermion Hamiltonian defined in Eqs. (2.22) and (7.7) with $\{q_i\}$ treated as a configuration of classical variables. The subscripts in these Hamiltonians indicate that the sign of the potential on that site is flipped relative to $\hat{H}(q)$.

The remaining free fermion correlators in Eq. (7.14) can be interpreted as the overlap of the two states

$$e^{i\hat{H}_{jk}(q)t}e^{-i\hat{H}(q)t}|\psi\rangle, \qquad \text{and} \qquad e^{i\hat{H}_{jklm}(q)t}e^{-i\hat{H}_{lm}(q)t}|\psi\rangle. \tag{7.15}$$

These states differ only in the sign of the potential on sites $l$ and $m$ in the time evolution. This overlap can then be measured using a similar interferometry experiment as above with spins on sites $l$ and $m$ which control the potential on those sites. The additional difficulty to the standard spin correlators is that in both of these states we have evolved forward and then backward again. We would therefore be required to first evolve with Hamiltonian $\hat{H}(q)$ for a time $t$, then change the Hamiltonian $\hat{H}(q) \rightarrow -\hat{H}_{jk}(q)$, and then evolve with this Hamiltonian for a further time $t$.

The main difficulties in the procedure lie with the process of changing $\hat{H}(q) \rightarrow -\hat{H}_{jk}(q)$, which must be done on sufficiently short time scales without excessively disturbing the system. The process consists of three main parts: (i) changing the sign of the potential on all sites except $j$ and $k$; (ii) flipping the control spins; (iii) changing the sign of the hopping for the fermion. Changing the sign of the potential on each site is the easiest of these three processes which can be implemented by modifying the laser configuration for the quantum gas microscope setup [5]. Flipping the control spins can be done using techniques discussed in Ref. [20] for selectively addressing single spins. Changing the sign of the hopping coefficients is then the remaining and possibly the most problematic step. This can be achieved using Floquet engineering [21], the downside of which is that it heats the system [22]. To establish experimental feasibility, the impact of all three of these processes must be taken into account. We note that a similar interferometry proposal using control spins was considered in Ref. [23]. Here the authors avoid flipping the Hamiltonian by instead coupling two copies of the system to the same control spins. The feasibility of our approach should therefore also be compared to this similar approach.

The possibility of simulating OTOCs for our model in this type of setup is exciting. There currently exist very few measurements of OTOCs in experiments [24, 25]. With this type of experiment it should be possible to access large systems in two dimensions. Interacting generalization of our model can be easily accessed without significant extra experimental difficultly, and the free fermion limit allows for reliable and efficient numerical benchmarking.

### 7.5.2 Other Experiments

While we have presented a protocol for a cold atomic gas experiment, it is not the only setting in which our model could be simulated. For example, the trapped ion experiment that studied the lattice Schwinger model in Ref. [4] could also be used for our model in 1D. The benefit of such experiments would be that the charges could be treated as quantum variables, rather than classical ones that need averaging over. To make a more direct connection to these experiments let us perform a Jordan-Wigner transformation on our spinless fermions to cast the Hamiltonian (2.22) in the form of the spin Hamiltonian

$$\hat{H} = -J\sum_j \left(\hat{\sigma}_j^x\hat{\sigma}_{j+1}^x + \hat{\sigma}_j^y\hat{\sigma}_{j+1}^y\right) + 2h\sum_j \hat{\mu}_j^z\hat{\sigma}_j^z, \tag{7.16}$$

where we have also written $\hat{q}_j = \hat{\mu}_j^z$ as a spin operator. Since this Hamiltonian only contains two point interactions between spin operators, the Trotterized version of the time evolution can be implemented in a similar trapped ion experiment, using local unitary operations and the Mølmer-Sørensen gates, as explained in Sect. 1.6.2.

There are other settings that could potentially be used to simulate our model that we have not yet fully investigated. One additional example that we will mention is quantum simulation on superconducting chips [1]. The basic elements of these systems are circuit-QED qubits, which are essentially LC-oscillators with Josephson junctions as non-linear inductors. When connected in an array, the system can be described by the lattice Jaynes-Cummings model, which is closely related to the Bose-Hubbard model.

## References

1. Houck AA, Türeci HE, Koch J (2012) On-chip quantum simulation with superconducting circuits. Nat Phys 8:292–299. https://doi.org/10.1038/nphys2251
2. Aspuru-Guzik A, Walther P (2012) Photonic quantum simulators. Nat Phys 8:285–291. https://doi.org/10.1038/nphys2253
3. Cirac JI, Zoller P (1995) Quantum computations with cold trapped ions. Phys Rev Lett 74:4091–4094. https://doi.org/10.1103/PhysRevLett.74.4091, arXiv:0305129 [quant-ph]

4. Martinez EA, Muschik CA, Schindler P, Nigg D, Erhard A, Heyl M, Hauke P, Dalmonte M, Monz T, Zoller P, Blatt R (2016) Real-time dynamics of lattice gauge theories with a few-qubit quantum computer. Nature 534:516–519. https://doi.org/10.1038/nature18318
5. Choi J-Y, Hild S, Zeiher J, Schauss P, Rubio-Abadal A, Yefsah T, Khemani V, Huse DA, Bloch I, Gross C (2016) Exploring the many-body localization transition in two dimensions. Science 352:1547–1552. https://doi.org/10.1126/science.aaf8834
6. Smith A, Knolle J, Kovrizhin DL, Moessner R (2017a) Disorder-free localization. Phys Rev Lett 118:266601. https://doi.org/10.1103/PhysRevLett.118.266601
7. Smith A, Knolle J, Moessner R, Kovrizhin DL (2017b) Absence of ergodicity without quenched disorder: from quantum disentangled liquids to many-body localization. Phys Rev Lett 119:176601. https://doi.org/10.1103/PhysRevLett.119.176601
8. Smith A, Knolle J, Moessner R, Kovrizhin DL (2018a) Dynamical localization in $Z_2$ lattice gauge theories. J Phys Rev B 97:245137
9. Fogarty JT, Gullo NL, Paternostro M, Busch T (2011) Orthogonality catastrophe as a consequence of qubit embedding in an ultracold Fermi gas. Phys Rev A 84:063632
10. Knap M, Shashi A, Nishida Y, Imambekov A, Abanin DA, Demler E (2012) Time-dependent impurity in ultracold fermions: orthogonality catastrophe and beyond. Phys Rev X 2:041020. https://doi.org/10.1103/PhysRevX.2.041020
11. Hangleiter D, Mitchison MT, Johnson TH, Bruderer M, Plenio MB, Jaksch D (2015) Nondestructive selective probing of phononic excitations in a cold Bose gas using impurities. Phys Rev A 91:013611
12. Buchleitner MA, Jaksch D, Mur-Petit J (2016) Measuring correlations of cold-atom systems using multiple quantum probes. Phys Rev A 94:053634
13. Montvay I, Münster G (1994) Quantum fields on a lattice. Cambridge University Press, Cambridge, UK. https://doi.org/10.1017/CBO9780511470783
14. Wiese U-J (2013) Ultracold quantum gases and lattice systems: quantum simulation of lattice gauge theories. Ann Phys 525:777–796
15. Zohar E, Cirac JI, Reznik B (2016) Quantum simulations of lattice gauge theories using ultracold atoms in optical lattices. Rep Prog Phys 79:014401. https://doi.org/10.1088/0034-4885/79/1/014401
16. Prosko C, Lee S-P, Maciejko J (2017) Simple $Z_2$ gauge theories at finite fermion density. Phys Rev B 96:205104. https://doi.org/10.1103/PhysRevB.96.205104
17. Smith A, Kovrizhin DL, Moessner R, Knolle J (2018b) Dynamics of a lattice gauge theory with fermionic matter minimal quantum simulator with time-dependent impurities in ultracold gases. Quantum Sci Technol 3:044003. https://doi.org/10.1088/2058-9565/aad39a, arXiv:1803.06575
18. Rüegg A, Huber SD, Sigrist M (2010) Z2-slave-spin theory for strongly correlated fermions. Phys Rev B 81:155118
19. Žitko R, Fabrizio M (2015) Z2 gauge theory description of the Mott transition in infinite dimensions. Phys Rev B 91:245130
20. Weitenberg C, Endres M, Sherson JF, Cheneau M, Schauß P, Fukuhara T, Bloch I, Kuhr S (2011) Single-spin addressing in an atomic Mott insulator. Nature 471:319–324. https://doi.org/10.1038/nature09827
21. Lignier H, Sias C, Ciampini D, Singh Y, Zenesini A, Morsch O, Arimondo E (2007) Dynamical control of matter-wave tunneling in periodic potentials. Phys Rev Lett 99:220403
22. Reitter M, Näger J, Wintersperger K, Sträter C, Bloch I, Eckardt A, Schneider U (2017) Interaction dependent heating and atom loss in a periodically driven optical lattice. Phys Rev Lett 119:200402
23. Yao NY, Grusdt F, Swingle B, Lukin MD, Stamper-Kurn DM, Moore JE, Demler EA (2016b) Interferometric approach to probing fast scrambling. arXiv:1607.01801
24. Garttner M, Bohnet JG, Safavi-Naini A, Wall ML, Bollinger JJ, Rey AM (2017) Measuring out-of-time-order correlations and multiple quantum spectra in a trapped-ion quantum magnet. Nat Phys 13:781–786. https://doi.org/10.1038/NPHYS4119
25. Li J, Fan R, Wang H, Ye B, Zeng B, Zhai H, Peng X, Du J (2017) Measuring out-of-time-order correlators on a nuclear magnetic resonance quantum simulator. Phys Rev X 7:031011

# Chapter 8
# Conclusions and Outlook

In this thesis we have addressed a long standing question about the dynamics of quantum systems—is quenched disorder a necessary requirement for localization? We have answered this question by introducing a family of translationally invariant models of coupled fermions and spins with a disorder-free mechanism for localization. The localization behaviour is analytically identified through an exact mapping to an Anderson localization problem, and is corroborated by large-scale numerical simulations of global quantum quenches.

We introduced the models in Chap. 2, where we gave details of the exact mapping to free fermions. Importantly, our model has an extensive set of conserved charges associated with the local $\mathbb{Z}_2$ gauge symmetry. These charges play the role of an effective binary potential for our fermions and our quench protocol amounts to averaging over these charge configurations. We also discussed the importance of the gauge symmetry and the connection to lattice gauge theories, in particular to the toric code.

In Chap. 3 we performed an extensive numerical study of dynamical correlators after a global quench. Due to the duality mapping, we were able to access system sizes much greater than the localization lengths using the free-fermion methods outlined in the Appendices. This numerical investigation is most crisp in one dimension where we studied quench protocols relevant to experiments, such as measuring remaining inhomogeneities after starting with a domain wall or charge density wave initial fermion configuration. The long-time distribution of densities and correlations are determined by the localization length of the single particle problem, which we are able to read off from these dynamical quantities.

We were also able to numerically study the dynamics of large systems in 2D, where many of the localization signatures carry over from 1D. The higher-dimensionality, however, provides an interesting new connection to a quantum site percolation problem. This connection is unique to the emergent binary disorder potential identified in our model, and is revealed in the strong effective disorder limit. Our numerical simulations are suggestive of a delocalization-by-percolation transition in 2D.



A. Smith, *Disorder-Free Localization*, Springer Theses,
https://doi.org/10.1007/978-3-030-20851-6_8

This further suggests the possibility of two localization transitions in 3D, one due to disorder, and the other due to percolation.

In Chap. 4 we studied the entanglement in our system following a quantum quench. Our investigation was motivated by the proposed quantum disentangled liquid (QDL) phase of matter and the projective measures of entanglement that are used to define it. These measures revealed a dependence on the choice of basis. They were unable to distinguish the localized fermions from the spin subsystem, despite the clear localization behaviour for the former that we are able to demonstrate using local observables. In the dual language, however, the long time state does indicate a QDL and the measure reveals the localization of the $c$-fermions. By deconstructing the definition of this entanglement measure we were able to find distinguishing behaviour, but further investigation is needed to understand this properly and to suggest a more conclusive signature of QDL. We also considered the quantum mutual information, which was able to distinguish the two subsystems, but we are yet to provide a complete interpretation of these results.

To study information spreading, we numerically simulated the out-of-time-ordered correlators for our gauge field in Chap. 5, which revealed a rich phenomenology. All of the OTOCs that we considered for the gauge field could be reduced to a "double Loschmidt echo" for free fermions averaged over disorder. Notably, we found logarithmic spreading of correlations, even when the localization length was an order of magnitude smaller than the system size. Similar and consistent behaviour was found in Ref. [1] for the standard Loschmidt echo in an Anderson localized system. In combination, these OTOCs demonstrate the complex dynamics in our system, despite the mapping to free fermions.

Disorder-free localization in our models relies heavily on the presence of conserved charges, which play the role of an effective potential. Despite this, there are many terms that we can add to the Hamiltonian that don't give dynamics to the charges, but do render the model interacting, which we consider in Chap. 6. In this setting we make a direct connection to the phenomenology of many-body localization. One of the characteristic properties of MBL is the logarithmic growth of entanglement following a quantum quench, which we also observe in the interacting extensions of our model. By adding a term that confines the spin excitations we also observe anomalously slow double logarithmic growth of entanglement, but in all cases we observe the persistent memory of initial states due to localization. We also numerically investigated the effect of perturbations that give dynamics to the charges. In this case our model reduces to a heavy-light particle mixture and we find a time scale after which localization is lost and the model becomes ergodic.

Finally, in Chap. 7 we propose an experiment to simulate the dynamics of the spin degrees of freedom in a two dimensional version of our model. These spins correspond to a gauge field and our model provides a minimal example of $\mathbb{Z}_2$ lattice gauge theory coupled to fermionic matter. We propose an experiment using cold atoms and a protocol for measuring dynamical gauge field correlations that is feasible with current experimental capabilities. We also extended this proposal to study out-of-time-ordered correlators, for which we give some details in Sect. 7.5.1. The mapping to free-fermions can be used to numerically benchmark the experiments but the additional interacting perturbations can also be included in the experiments.

There are many open questions that we have not addressed in this thesis. We will mention a few of them here that we have already alluded to in the main text. For example, in Sect. 2.3.1 we introduced an extension to our model that is closely related to the Hubbard model. We considered adding a second species of fermions to our model, corresponding to the opposite spin orientation, and the Hamiltonian takes the form

$$\hat{H} = -J \sum_{j,\alpha=\uparrow,\downarrow} \hat{\sigma}^z_{j,j+1} \left( \hat{f}^\dagger_{j,\alpha} \hat{f}_{j+1,\alpha} + \text{H.c.} \right) - h \sum_j \hat{\sigma}^x_{j-1,j} \hat{\sigma}^x_{j,j+1}, \tag{8.1}$$

which also has an extensive set of conserved quantities $\hat{q}_j = \hat{\sigma}^x_{j-1,j} \hat{\sigma}^x_{j,j+1} (-1)^{\hat{n}_{j\uparrow}+\hat{n}_{j\downarrow}}$. After a duality transformation of the spins, the Hamiltonian can be rewritten as

$$\hat{H} = -J \sum_{j,\alpha=\uparrow,\downarrow} \left( \hat{c}^\dagger_{j,\alpha} \hat{c}_{j+1,\alpha} + \text{H.c.} \right) - 4h \sum_j \hat{q}_j \hat{n}_{j,\uparrow} \hat{n}_{j,\downarrow} + 2h \sum_j \hat{q}_j (\hat{n}_{j,\uparrow} + \hat{n}_{j,\downarrow}) - h \sum_j \hat{q}_j, \tag{8.2}$$

which is in the form of the Hubbard model with potential and interactions determined by the conserved charges $q$. The Hubbard model is a paradigmatic model of strongly correlated electrons and so this extension of our model is a particularly interesting direction of future research. This model can also be simulated using an extension of the experimental protocol that we have outlined in Chap. 7. A constrained version of this model was considered in Refs. [2, 3], which allowed them to map out the ground state phase diagram for the Hubbard model. We hope that the Hamiltonian (8.1) may also aid the understanding of dynamics of the Hubbard model.

As a second example, the entanglement properties of our model also warrant further investigation. Due to the difficulties of defining a meaningful measure of tripartite entanglement, our model provides an important example that can in principle be understood in terms of free fermions. Even here it remains non-trivial to disentangle the degrees of freedom and to discuss the entanglement of a particular subsystem. In this thesis we have made some progress in uncovering the properties of the projective measure of entanglement used to define the QDL, but our understanding is still incomplete. Further, we have shown that the quantum mutual information may provide an alternative diagnosis for the QDL. We suggest that, for a more complete picture, other entanglement measures should also be considered. An example that has different properties to both of those considered in this thesis is the negativity [4–6], which can be used to distinguish the quantum and thermal entanglement in mixed states. Our model should provide a controlled setting for understanding the properties of this measure, which in turn may further shed light on the complex entanglement between the different degrees of freedom in our model.

Although we have demonstrated for the first time that complete localization is possible in a disorder-free model, it remains inconclusive whether a disorder-free mechanism for localization can be robust to generic perturbations. We think some of the most promising suggestions in this direction are quantum analogues of classically glassy models. In Ref. [7], the authors consider the quantum East model and find behaviour consistent with MBL for all times accessible by numerics. However, the authors were unable to conclude that this behaviour continues indefinitely in the thermodynamic limit. Furthermore, since these models rely on kinetic constraints, it is unclear to what extent these can be robust to generic perturbations. These questions are further compounded by the evidence that MBL induced by quenched disorder is also unstable in higher than one dimension [8, 9].

Finally, we are very excited about the prospect of realising a two-dimensional version of our system in experiments. We feel that our proposal will push the boundaries for the simulation of quantum lattice gauge theories, which are so far restricted to one dimension. Furthermore, our free fermion mapping provides a well controlled numerically accessible limit for benchmarking, about which the model can be perturbed. Particularly exciting is the prospect of experimentally measuring out-of-time-ordered correlators in one and two dimensions. Our proposed protocols are simple enough that they can be performed with current experimental capabilities, and as we have already shown in this thesis, provide access to non-trivial quantum dynamics, even in the free fermion limit.

While our understanding has advanced dramatically since the hypothesis of eigenstate thermalization and establishing the existence of the many-body localized phase, it is clear that there is still much left to understand about how quantum systems relax. In this thesis we have unveiled the rich phenomenology of non-equilibrium dynamics in a minimal setting. Thus, we have discovered a disorder-free mechanism of localization and drawn connections to paradigmatic models of condensed matter physics. In the future, we expect our tractable setting to shed further light on non-equilibrium quantum phenomena and the physics of lattice gauge theories.

## References

1. Vardhan S, De Tomasi G, Heyl M, Heller EJ, Pollmann F (2017) Characterizing time irreversibility in disordered fermionic systems by the effect of local perturbations. Phys Rev Lett 119:016802. https://doi.org/10.1103/PhysRevLett.119.016802
2. Rüegg A, Huber SD, Sigrist M (2010) Z2-slave-spin theory for strongly correlated fermions. Phys Rev B 81:155118. https://doi.org/10.1103/PhysRevB.81.155118
3. Žitko R, Fabrizio M (2015) Z2 gauge theory description of the Mott transition in infinite dimensions. Phys Rev B 91:245130. https://doi.org/10.1103/PhysRevB.91.245130
4. Życzkowski K, Horodecki P, Sanpera A, Lewenstein M (1998) Volume of the set of separable states. Phys Rev A 58:883–892. https://doi.org/10.1103/PhysRevA.58.883
5. Eisert J (2001) Entanglement in quantum information theory. PhD thesis
6. Vidal G, Werner RF (2002) Computable measure of entanglement. Phys Rev A 65:032314. https://doi.org/10.1103/PhysRevA.65.032314
7. van Horssen M, Levi E, Garrahan JP (2015) Dynamics of many-body localization in a translation-invariant quantum glass model. Phys Rev B 92:100305. https://doi.org/10.1103/PhysRevB.92.100305
8. De Roeck W, Huveneers F (2017) Stability and instability towards delocalization in many-body localization systems. Phys Rev B 95:155129. https://doi.org/10.1103/PhysRevB.95.155129
9. Potirniche I-D, Banerjee S, Altman E (2018) On the stability of many-body localization in $d > 1$. arXiv:1805.01475

# Appendix A
# Free Fermion Correlators

In this appendix we will calculate the connected density-density correlator

$$C_{lm}(t) = \langle\psi(t)|\,\hat{n}_l\hat{n}_m\,|\psi(t)\rangle_c = \langle\psi(t)|\,\hat{n}_l\hat{n}_m\,|\psi(t)\rangle - \langle\psi(t)|\,\hat{n}_l\,|\psi(t)\rangle\langle\psi(t)|\,\hat{n}_m\,|\psi(t)\rangle, \tag{A.1}$$

where evolution is determined by the free fermion Hamiltonian

$$\hat{H} = -\sum_{j=1}^{N}(\hat{c}_j^\dagger\hat{c}_{j+1} + \hat{c}_{j+1}^\dagger\hat{c}_j), \tag{A.2}$$

with periodic boundary conditions. The initial state for the fermions that we consider is the density-wave pattern $|\psi(0)\rangle = |\cdots 101010\cdots\rangle$, with fermions on odd sites only. We choose to work in the Heisenberg representation with stationary states and evolving operators. In this representation the correlator is given by

$$C_{lm}(t) = \langle\psi|\,\hat{n}_l(t)\hat{n}_m(t)\,|\psi\rangle_c = \langle\psi|\,\hat{n}_l(t)\hat{n}_m(t)\,|\psi\rangle - \langle\psi|\,\hat{n}_l(t)\,|\psi\rangle\langle\psi|\,\hat{n}_m(t)\,|\psi\rangle, \tag{A.3}$$

with $|\psi\rangle = |\psi(0)\rangle$. In this appendix we derive the analytic results of Eqs. (1.4) and (1.7).

## A.1 Diagonalising the Hamiltonian

Due to translation invariance of the Hamiltonian (A.2) it can be diagonalized by Fourier transform, i.e.,

$$\hat{c}_j = \frac{1}{\sqrt{N}}\sum_k e^{ikj}\hat{c}_k, \qquad \hat{c}_k = \frac{1}{\sqrt{N}}\sum_j e^{-ikj}\hat{c}_j, \tag{A.4}$$


A. Smith, *Disorder-Free Localization*, Springer Theses,
https://doi.org/10.1007/978-3-030-20851-6

where the lattice momenta $k = 2\pi n/N$ with $n \in \{0, \ldots, N-1\}$. We will distinguish between the different operators on either side of the Fourier transform by only using $k, q, k', q'$ to label the lattice momenta. This transforms the Hamiltonian (A.2) to the diagonal form

$$\hat{H} = \sum_k \varepsilon_k \hat{c}_k^\dagger \hat{c}_k, \tag{A.5}$$

where $\varepsilon_k = -2\cos k$. The time dependence of the operators is found via the Heisenberg equation of motion

$$\frac{d}{dt}\hat{c}_k(t) = i[\hat{H}, \hat{c}_k] = -i\varepsilon_k \hat{c}_k(t) \quad \Rightarrow \quad \hat{c}_k(t) = e^{-i\varepsilon_k t}\hat{c}_k, \tag{A.6}$$

and the time dependence in the position representation is given by

$$\hat{c}_j(t) = \frac{1}{\sqrt{N}} \sum_k e^{ikj} e^{-i\varepsilon_k t} \hat{c}_k. \tag{A.7}$$

We now proceed to calculate the average on-site density $\langle\psi|\hat{n}_l(t)|\psi\rangle$ and the density correlator $\langle\psi|\hat{n}_l(t)\hat{n}_m(t)|\psi\rangle$.

## A.2 Average Density Correlators

We begin by calculating the average density. We first rewrite the time dependence of the density operator using Eq. (A.7), i.e.,

$$\langle\psi|\hat{n}_l(t)|\psi\rangle = \frac{1}{N} \sum_{k,k'} e^{-i(k-k')l} e^{i(\varepsilon_k - \varepsilon_{k'})t} \langle\psi|\hat{c}_k^\dagger \hat{c}_{k'}|\psi\rangle. \tag{A.8}$$

The remaining expectation value on the right-hand side no longer contains any time dependence and we Fourier transform it to position space. Doing so we find

$$\langle\psi|\hat{c}_k^\dagger \hat{c}_{k'}|\psi\rangle = \frac{1}{N} \sum_{\alpha,\beta} e^{-ik\alpha} e^{-ik'\beta} \langle\psi|\hat{c}_\alpha^\dagger \hat{c}_\beta|\psi\rangle. \tag{A.9}$$

The inner products appearing on the right-hand side are now particularly simple, and are equal to one when $\alpha = \beta$ and odd, and zero otherwise, due to the form of the initial state. Thus, we have

$$\langle\psi|\hat{c}_k^\dagger \hat{c}_{k'}|\psi\rangle = \frac{1}{2} e^{i(k-k')} \frac{1}{N/2} \sum_{\alpha=1}^{N/2} e^{2i(k-k')\alpha} = \frac{1}{2}(\delta_{k,k'} - \delta_{k,k'+\pi}), \tag{A.10}$$

where we note that $N$ must be even for the charge density wave to be consistent with the periodic boundary conditions. Plugging Eq. (A.10) into Eq. (A.8) we get the expression for the density expectation value,

$$\langle\psi|\hat{n}_l(t)|\psi\rangle = \frac{1}{2} - \frac{(-1)^l}{2N}\sum_k e^{2i\varepsilon_k t}. \tag{A.11}$$

## A.3 Density-Density Correlators

Next consider the density-density correlator. First we pull out the time dependence of the operators, to get

$$\langle\psi|\hat{n}_l(t)\hat{n}_m(t)|\psi\rangle = \frac{1}{N^2}\sum_{k,k',q,q'} e^{-i(k-k')l-i(q-q')m} e^{i(\varepsilon_k-\varepsilon_{k'})t+i(\varepsilon_q-\varepsilon_{q'})t}\langle\psi|\hat{c}_k^\dagger\hat{c}_{k'}\hat{c}_q^\dagger\hat{c}_{q'}|\psi\rangle. \tag{A.12}$$

We then compute the remaining expectation values by moving to position space using the Fourier transform, to get

$$\langle\psi|\hat{c}_k^\dagger\hat{c}_{k'}\hat{c}_q^\dagger\hat{c}_{q'}|\psi\rangle = \frac{1}{N^2}\sum_{\alpha,\beta,\gamma,\delta} e^{ik\alpha}e^{-ik'\beta}e^{iq\gamma}e^{-iq'\delta}\langle\psi|\hat{c}_\alpha^\dagger\hat{c}_\beta\hat{c}_\gamma^\dagger\hat{c}_\delta|\psi\rangle. \tag{A.13}$$

The remaining expectation value on the right-hand side is equal to one if $\delta=\gamma$, $\alpha=\beta$ and all indices are odd, or if $\alpha=\delta$ odd and $\beta=\gamma$ even, and zero otherwise. We can write this in the shorthand notation $\langle\psi|\hat{c}_\alpha^\dagger\hat{c}_\beta\hat{c}_\gamma^\dagger\hat{c}_\delta|\psi\rangle = \delta_{\alpha,\beta-\text{odd}}\delta_{\gamma,\delta-\text{odd}} + \delta_{\alpha,\delta-\text{odd}}\delta_{\beta,\gamma-\text{even}}$. Equation (A.13) can then be written as

$$\frac{1}{N^2}\sum_{\alpha,\beta-\text{odd}} e^{i(k-k')\alpha}e^{i(q-q')\beta} + \frac{1}{N^2}\sum_{\alpha-\text{odd}}\sum_{\beta-\text{even}} e^{i(k-q')\alpha}e^{i(q-k')\beta}. \tag{A.14}$$

These sums factorise and Eq. (A.14) is simplified by using the following identity:

$$\sum_{\alpha-\text{odd}} e^{ik\alpha} = e^{-ik}\sum_{\alpha=1}^{N/2} e^{2ik\alpha} = e^{-ik}\frac{N}{2}(\delta_{k,0}+\delta_{k,\pi}) = \frac{N}{2}(\delta_{k,0}-\delta_{k,\pi}). \tag{A.15}$$

In total, the density correlator (A.12) is given by

$$\langle\psi|\hat{n}_l(t)\hat{n}_m(t)|\psi\rangle = \frac{1}{4N^2}\sum_{k,k',q,q'} e^{-i(k-k')l}e^{-i(q-q')m}e^{i(\varepsilon_k-\varepsilon_{k'})t}e^{i(\varepsilon_q-\varepsilon_{q'})t} \times\left[(\delta_{k,k'}-\delta_{k,k'+\pi})(\delta_{q,q'}-\delta_{q,q'+\pi}) + (\delta_{k,q'}-\delta_{k,q'+\pi})(\delta_{q,k'}+\delta_{q,k'+\pi})\right]. \tag{A.16}$$

Performing the sums over $k'$ and $q'$, the first half of the square bracket gives

$$\left(\frac{1}{2}-\frac{(-1)^l}{2N}\sum_k e^{2i\varepsilon_k t}\right)\left(\frac{1}{2}-\frac{(-1)^m}{2N}\sum_k e^{2i\varepsilon_q t}\right) \tag{A.17}$$

which cancels the term $\langle\Psi|n_l(t)|\Psi\rangle\langle\Psi|n_m(t)|\Psi\rangle$ in the connected correlator (A.3). Expanding the second half of the square bracket we get 4 terms which we will work through one-by-one. Firstly the $\delta_{k,q'}\delta_{q,k'}$ term gives

$$\frac{1}{4N^2}\left|\sum_k e^{-ik(l-m)}\right|^2 = \frac{1}{4}\delta_{l,m}. \tag{A.18}$$

Next, we look at the $\delta_{k,q'}\delta_{q,k'+\pi}$ term which gives

$$\frac{(-1)^l}{4N^2}\sum_k e^{-ik(l-m)}\sum_q e^{iq(l-m)}e^{2i\varepsilon_q t} = \frac{(-1)^l}{4N}\delta_{l,m}\sum_q e^{2i\varepsilon_q t}, \tag{A.19}$$

where we used the fact $\varepsilon_{k-\pi} = -\varepsilon_k$. Next, the $-\delta_{k,q'+\pi}\delta_{q,k'}$ term gives

$$-\frac{(-1)^m}{4N^2}\sum_q e^{iq(l-m)}\sum_k e^{-ik(l-m)}e^{2i\varepsilon_k t} = -\frac{(-1)^m}{4N}\delta_{l,m}\sum_k e^{2i\varepsilon_k t}, \tag{A.20}$$

which cancels the previous term. Finally, we consider the $-\delta_{k,q'+\pi}\delta_{q,k'+\pi}$ term which is

$$-\frac{(-1)^{(l-m)}}{4}\left(\frac{1}{N}\sum_q e^{iq(l-m)}e^{-4i\cos(q)t}\right)^2 \tag{A.21}$$

Putting this all together we arrive at

$$\langle\psi|\hat{n}_l(t)\hat{n}_m(t)|\psi\rangle_c = \frac{1}{4}\delta_{l,m} - \frac{(-1)^{(l-m)}}{4}\left(\frac{1}{N}\sum_q e^{iq(l-m)}e^{-4i\cos(q)t}\right)^2. \tag{A.22}$$

## A.4 Thermodynamic Limit

Our expressions for the density average and density-density correlator involve sums of the form

$$\frac{1}{N}\sum_k e^{ik(l-m)}e^{-4i\cos(k)t}. \tag{A.23}$$

It turns out that we can obtain an explicit closed form expression for these sums in the thermodynamic limit $N \to \infty$. The lattice momenta take values $k = 2\pi n/N$ for $n \in \{0, \dots, N-1\}$, and so we can rewrite the sum as

$$\frac{1}{N}\sum_{n=0}^{N-1} e^{2i\pi n/N(l-m)} e^{-4i\cos(2\pi n/N)t}. \tag{A.24}$$

Taking the limit $N \to \infty$ we get

$$\frac{2\pi n}{N} \to x \in [0, 2\pi), \quad \text{and} \quad \frac{1}{N}\sum_{n=0}^{N-1} \to \frac{1}{2\pi}\int_0^{2\pi} \mathrm{d}x, \tag{A.25}$$

and thus in this limit we find

$$\frac{1}{N}\sum_k e^{ik(l-m)} e^{2i\varepsilon_k t} \xrightarrow[N\to\infty]{} \frac{1}{2\pi}\int_0^{2\pi} \mathrm{d}x \; e^{ix(l-m)-4i\cos(x)t} = i^{(l-m)} J_{l-m}(4t), \tag{A.26}$$

where $J_\alpha(x)$ are the Bessel functions of the first kind.

In this limit the average density (A.11) is therefore given by

$$\langle\psi|\hat{n}_l(t)|\psi\rangle = \frac{1}{2} - \frac{(-1)^l}{2} J_0(4t). \tag{A.27}$$

This further means that the average density imbalance of Eq. (1.7) is given by

$$\Delta\rho(t) = \frac{1}{N}\sum_j |\langle\psi|\hat{n}_j(t) - \hat{n}_{j+1}(t)|\psi\rangle| = |J_0(4t)|, \tag{A.28}$$

as shown in Fig. 1.1b.

The connected density correlator of Eq. (1.4) is given by

$$\langle\Psi|n_l(t)n_m(t)|\Psi\rangle_c = \frac{1}{4}\delta_{l,m} - \frac{1}{4}J_{m-l}(4t)^2, \tag{A.29}$$

and is shown in Fig. 1.1a.

# Appendix B
# Calculation of Fermion Correlators Using Determinants

In the case of a Hamiltonian that is bilinear in fermion operators, dynamic correlation functions can be obtained in terms of determinants of single particle matrices. In the main text we have shown that general correlators for our family of models can be written in terms of purely fermionic correlators. In the following we explain how the calculation of these correlators can be reduced to determinants, see e.g. [1]. A mapping to a free-fermion Hamiltonian dramatically decreases the computational cost compared with ED, which allows us to reach much larger system sizes.

## B.1 Derivation of the Determinant Expressions

Generically, we are interested in computing expressions of the form

$$\langle\alpha|\exp\{i\sum_{ij}A_{ij}\hat{c}_i^\dagger\hat{c}_j\}|\beta\rangle, \tag{B.1}$$

where $A$ is a Hermitian matrix, and $|\alpha\rangle = \hat{c}_{m_M}^\dagger\cdots\hat{c}_{m_1}^\dagger|vac\rangle$ and $|\beta\rangle = \hat{c}_{n_M}^\dagger\cdots\hat{c}_{n_1}^\dagger|vac\rangle$ are fermionic Slater determinants, with $M \leq N$, where $N$ is the number of single particle states. To proceed with the calculation of (B.1) we first use the unitarity of the exponential operator $\hat{U}_A \equiv \exp\{i\sum_{ij}A_{ij}\hat{c}_i^\dagger\hat{c}_j\}$ to rewrite (B.1) as

$$\langle vac|\hat{c}_{m_1}\cdots\hat{c}_{m_N}\hat{U}_A\hat{c}_{n_N}^\dagger\hat{U}_A^\dagger\cdots\hat{U}_A\hat{c}_{n_1}^\dagger\hat{U}_A^\dagger\hat{U}_A|vac\rangle = \langle vac|\hat{c}_{m_1}\cdots\hat{c}_{m_N}\hat{\tilde{c}}_{n_N}^\dagger\cdots\hat{\tilde{c}}_{n_1}^\dagger|vac\rangle, \tag{B.2}$$

where $\hat{\tilde{c}}_j^\dagger \equiv \hat{U}_A\hat{c}_j^\dagger\hat{U}_A^\dagger$ and we have used that $\hat{U}_A|vac\rangle = |vac\rangle$. With the help of the Baker-Campbell-Hausdorff formula we obtain

$$\hat{\tilde{c}}_i^\dagger = \sum_j \exp\{iA^T\}_{ij}\hat{c}_j^\dagger \equiv \sum_j U_{A,ij}^T\hat{c}_j^\dagger, \tag{B.3}$$



A. Smith, *Disorder-Free Localization*, Springer Theses,
https://doi.org/10.1007/978-3-030-20851-6

distinguishing between the operator $\hat{U}_A$ and the matrix $U_A$ by a hat. Finally, we insert (B.3) into (B.2), and use the fermionic anti-commutation relations to obtain

$$\langle\alpha|\hat{U}_A|\beta\rangle = \det D, \quad D_{jk} = [U_A]_{n_j m_k}, \tag{B.4}$$

with $j, k = 1, \ldots, M$. In other words we select from the matrix $U_A$ those rows and columns that correspond to occupied states in Slater determinants $|\beta\rangle$, $|\alpha\rangle$. For example, if $|\alpha\rangle = |\beta\rangle$ is a 1D CDW with fermions on the odd sites, then the matrix $D$ is made by taking only the odd rows and columns of $U_A$.

This derivation allows for further generalisations. For example, in the case of an arbitrary number of unitary operators, using repeatedly the Baker-Campbell-Hausdorff formula we obtain

$$\langle\alpha|\hat{U}_A\hat{U}_B\cdots|\beta\rangle = \det D, \quad D_{jk} = [U_A U_B\cdots]_{n_j m_k}. \tag{B.5}$$

This equation is suitable for evaluating correlators similar to the one in Eq. (2.23). In case of the fermion correlators we need to consider expressions of the following form

$$C_{kl} = \langle\alpha|\hat{c}_k^\dagger \exp\{i\sum_{ij} A_{ij}\hat{c}_i^\dagger\hat{c}_j\}\hat{c}_l|\beta\rangle. \tag{B.6}$$

By commuting $\hat{c}_k^\dagger$ to the left, and $\hat{c}_l$ to the right, we pick up factors $(-1)^{N-p}$ and $(-1)^{N-q}$, where $m_p = k$ and $n_q = l$ and arrive at

$$\begin{aligned} C_{kl} = (-1)^{p+q}\langle vac|\hat{c}_{m_1}\cdots\hat{c}_{m_{p-1}}\hat{c}_{m_{p+1}}\cdots\hat{c}_{m_N} \\ \times\hat{U}_A\hat{c}_{n_N}^\dagger\cdots\hat{c}_{n_{q+1}}^\dagger\hat{c}_{n_{q-1}}^\dagger\cdots\hat{c}_{n_1}^\dagger|vac\rangle. \end{aligned} \tag{B.7}$$

In this case we need to remove the $q$-row and the $p$-column from the matrix $D$ before taking the determinant and then multiply by the corresponding sign. Specifically we need the $q-p$ cofactor of $D$, where $D$ is given in Eq. (B.4). The final expression of the fermion correlator (B.6) now can be written in a simple form

$$C_{kl} = D_{lk}^{-1}\det D, \tag{B.8}$$

where $D_{jk} = [U_A]_{n_j m_k}$, $j, k = 1, \ldots M$.

The free-fermion mapping presented in the main text allows one to extract dynamical correlators for system sizes far beyond exact diagonalization. We can estimate the size of the fermionic Hilbert space at half-filling as $N^{-1/2}2^N$ with the spin degrees of freedom adding another factor of $2^N$. Instead of diagonalizing exponentially large matrices the identification of conserved charges allows us to sample uniformly from $\sim 2^N$ determinants of $N\times N$ matrices, corresponding to different charge configurations. Finally, finite-size scaling, as well as exact results (up to $N = 20$), show that the required number of samples for a given accuracy scales polynomially with $N$. Typically we sample over $10^3$–$10^4$ charge configurations.

## B.2 Examples

The expression for fermion correlators in terms of determinants that were derived in the previous section are sufficient for us to calculate all correlators used in the main text. In this section we will go through each of the correlators that we use and explicitly derive their expression in terms of single-particle matrices and determinants. We will assume the initial state (which is the same as the final) has definite site occupation numbers, i.e. $|\psi\rangle = \hat{c}^\dagger_{n_M} \cdots \hat{c}_{n_1}|vac\rangle$, where $M = N_f$ the fermion filling, $n_i \in \{1, \dots N\}$, $N$ is the number of sites, and $\hat{c}^\dagger_{n_i}$ creates a fermion at site $n_i$. For fixed particle number, the initial state can always be considered as having definite occupation in some basis, and we will consider a change of basis in Sect. B.3.

### B.2.1 Spin Correlators and OTOCs

In the main text we consider the average magnetization of the $z$-component of spin, the two-point equal time correlator for the $z$-component, and the out-of-time-ordered correlators for all combinations of $x$ and $z$ Pauli operators. They can all be written as averages over fermion correlators, and explicitly they take the form

$$\langle\Psi|\hat{\sigma}^z_{jk}(t)|\Psi\rangle = \frac{1}{2^{N-1}} \sum_{\{q_i\}=\pm1}{}' \langle\psi|e^{i\hat{H}(q)t}e^{-i\hat{H}_{jk}(q)t}|\psi\rangle \tag{B.9a}$$

$$\langle\Psi|\hat{\sigma}^z_{jk}(t)\hat{\sigma}^z_{lm}(t)|\Psi\rangle = \frac{1}{2^{N-1}} \sum_{\{q_i\}=\pm1}{}' \langle\psi|e^{i\hat{H}(q)t}e^{-i\hat{H}_{jklm}(q)t}|\psi\rangle \tag{B.9b}$$

$$\langle\Psi|\hat{\sigma}^\alpha_{jk}(t)\hat{\sigma}^\beta_{lm}\hat{\sigma}^\alpha_{jk}(t)\hat{\sigma}^\beta_{lm}|\Psi\rangle = \frac{1}{2^{N-1}} \sum_{\{q_i\}=\pm1}{}' \langle\psi|e^{i\hat{H}(q)t}e^{-i\hat{H}^{(\alpha)}_{jk}(q)t}e^{i\hat{H}^{(\alpha,\beta)}_{jk,lm}(q)t}e^{-i\hat{H}^{(\beta)}_{lm}(q)t}|\psi\rangle, \tag{B.9c}$$

where $\hat{H}(q)$ is defined in Eq. (2.22). The subscripts in Eq. (B.9a–b) indicate that the potential on the site is flipped, and for the OTOC in Eq. (B.9c) the subscripts indicate flipped potentials if the indices correspond to $\hat{\sigma}^z$ and a flipped hopping if they correspond to $\hat{\sigma}^x$, see Eq. (5.6).

In all four cases the fermion correlators are of the form $\langle\psi|e^{\hat{A}}e^{\hat{B}}\cdots|\psi\rangle = \det[U_A U_B \cdots]_{n_i n_i}$, where $U_A = \exp\{A\}$, and $A$ is the single particle matrix representation of the operator $\hat{A}$. The subscripts indicate that we take the determinant of sub-matrix $[U_A U_B \cdots]_{n_i n_i}$, where $n_i$ indices correspond to filled sites. We will consider the more general case in Sect. B.3. Explicitly, this gives

$$\langle\psi|e^{i\hat{H}(q)t}e^{-i\hat{H}_{jk}(q)t}|\psi\rangle = \det[U_H^\dagger U_{H_{jk}}]_{n_i n_i} \tag{B.10a}$$

$$\langle\psi|e^{i\hat{H}(q)t}e^{-i\hat{H}_{jklm}(q)t}|\psi\rangle = \det[U_H^\dagger U_{H_{jklm}}]_{n_i n_i} \tag{B.10b}$$

$$\langle\psi|e^{i\hat{H}(q)t}e^{-i\hat{H}^{(\alpha)}_{jk}(q)t}e^{i\hat{H}^{(\alpha,\beta)}_{jk,lm}(q)t}e^{-i\hat{H}^{(\beta)}_{lm}(q)t}|\psi\rangle = \det\left[U_H^\dagger U_{H^{(\alpha)}_{jk}}U^\dagger_{H^{(\alpha,\beta)}_{jk,lm}}U_{H^{(\beta)}_{lm}}\right]_{n_i n_i}, \tag{B.10c}$$

where $U_H = \exp\{-iHt\}$, $U_{H_{jk}} = \exp\{-iH_{jk}t\}$, etc.

### B.2.2 Average Density

When calculating fermion correlators we need to cast them in the form of Eq. (B.6). Let us first consider the average on-site density

$$\langle\Psi|\hat{n}_j(t)|\Psi\rangle = \frac{1}{2^{N-1}}\sum_{\{q_i\}=\pm 1}{}' \langle\psi|e^{i\hat{H}(q)t}\hat{c}_j^\dagger\hat{c}_j e^{-i\hat{H}(q)t}|\psi\rangle, \tag{B.11}$$

where in this case there are no flipped potentials between forwards and backwards evolution since $\hat{f}_j^\dagger\hat{f}_j = \hat{c}_j^\dagger\hat{c}_j$. To get this in the correct form we need to commute $\hat{c}_j^\dagger$ to the left and $\hat{c}_j$ to the right. Using the Baker-Campbell-Haussdorf formula we have that

$$\hat{U}^\dagger\hat{c}_j\hat{U} = \sum_k U_{jk}\hat{c}_k, \qquad \hat{U}^\dagger\hat{c}_j^\dagger\hat{U} = \sum_k U_{jk}^*\hat{c}_k^\dagger, \tag{B.12}$$

where $\hat{U} = \exp\{-i\hat{H}t\}$ and $U = \exp\{-iHt\}$, where $\hat{H}$ is a Hamiltonian operator and $H$ is the corresponding single-particle matrix representation. Using these expressions we get

$$\langle\psi|e^{i\hat{H}(q)t}\hat{c}_j^\dagger\hat{c}_j e^{-i\hat{H}(q)t}|\psi\rangle = \sum_{lm} U_{jl}^* U_{jm}\langle\psi|\hat{c}_l^\dagger e^{i\hat{H}(q)t}e^{-i\hat{H}(q)t}\hat{c}_m|\psi\rangle. \tag{B.13}$$

Using that $\hat{U}^\dagger\hat{U} = \mathbb{1}$, i.e., that $\hat{U}$ is unitary we find that the remaining fermion correlator takes the form $\langle\psi|\hat{c}_l^\dagger\hat{c}_m|\psi\rangle = \delta_{lm}\sum_q \delta_{ln_q}$, where $n_q$ are the indices corresponding to filled single particle basis states in $|\psi\rangle$. Finally, we find the neat form

$$\langle\psi|e^{i\hat{H}(q)t}\hat{c}_j^\dagger\hat{c}_j e^{-i\hat{H}(q)t}|\psi\rangle = \sum_{n_q} U_{jn_q}U^\dagger_{n_q j}, \tag{B.14}$$

for the average density for each charge configuration. Note that we are not required to compute a determinant and so density averages can be computed extremely efficiently.

### *B.2.3 Density Correlators*

Finally, let us consider the density correlator

$$\langle\Psi|\hat{n}_j(t)\hat{n}_k(t)|\Psi\rangle = \frac{1}{2^{N-1}}\sum_{\{q_i\}=\pm1}{}' \langle\psi|e^{i\hat{H}(q)t}\hat{c}_j^\dagger\hat{c}_j\hat{c}_k^\dagger\hat{c}_k e^{-i\hat{H}(q)t}|\psi\rangle. \tag{B.15}$$

To proceed we first need to get all creation operators on the right, annihilation operators on the left, and the unitary evolution operators in the middle. The first step is to use the anti-commutation relations for the fermion operators to get

$$\begin{aligned}&\langle\psi|e^{i\hat{H}(q)t}\hat{c}_j^\dagger\hat{c}_j\hat{c}_k^\dagger\hat{c}_k e^{-i\hat{H}(q)t}|\psi\rangle\\ &\quad= \delta_{jk}\langle\psi|e^{i\hat{H}(q)t}\hat{c}_j^\dagger\hat{c}_j e^{-i\hat{H}(q)t}|\psi\rangle + (1-\delta_{jk})\langle\psi|e^{i\hat{H}(q)t}\hat{c}_j^\dagger\hat{c}_k^\dagger\hat{c}_k\hat{c}_j e^{-i\hat{H}(q)t}|\psi\rangle,\end{aligned} \tag{B.16}$$

where the first term is the same as for the average density covered in the previous section, and so we will focus on the second term. Again we make use of the Baker-Campbell-Haussdorf formula to get

$$\langle\psi|e^{i\hat{H}(q)t}\hat{c}_j^\dagger\hat{c}_k^\dagger\hat{c}_k\hat{c}_j e^{-i\hat{H}(q)t}|\psi\rangle = \sum_{l,m,r,s} U_{jl}^* U_{ks}^* U_{kr} U_{jm}\langle\psi|\hat{c}_l^\dagger\hat{c}_s^\dagger\hat{c}_r\hat{c}_m|\psi\rangle, \tag{B.17}$$

where the unitary operators have cancelled since $\hat{U}^\dagger\hat{U} = \mathbb{1}$. Finally, this fermion correlator is zero unless $l = m$ and $s = r$, or $l = r$ and $s = m$. In both cases we need that all indices correspond to filled states in the state $|\psi\rangle$. Furthermore, we pick up a minus sign in the second case. This can be written as $\langle\psi|\hat{c}_l^\dagger\hat{c}_s^\dagger\hat{c}_r\hat{c}_m|\psi\rangle = (\delta_{lm}\delta_{rs} - \delta_{lr}\delta_{sm})\sum_{n_q,n_p}\delta_{ln_q}\delta_{sn_p}$. In total we find

$$\langle\psi|e^{i\hat{H}(q)t}\hat{c}_j^\dagger\hat{c}_k^\dagger\hat{c}_k\hat{c}_j e^{-i\hat{H}(q)t}|\psi\rangle = \sum_{n_q,n_p}\left(U_{jn_q}U_{n_qj}^\dagger U_{kn_p}U_{n_pk}^\dagger - U_{jn_q}U_{n_qk}^\dagger U_{kn_p}U_{n_pj}^\dagger\right). \tag{B.18}$$

The first term is simply the product of the average density on-sites $j$ and $k$, whereas the second contains the connected correlations between them. If we consider the connected correlator, that is $\langle\hat{n}_j(t)\hat{n}_k(t)\rangle_c = \langle\hat{n}_j(t)\hat{n}_k(t)\rangle - \langle\hat{n}_j(t)\rangle\langle\hat{n}_k(t)\rangle$, then we find

$$\langle\Psi|\hat{n}_j(t)\hat{n}_k(t)|\Psi\rangle_c = \frac{1}{2^{N-1}}\sum_{\{q_i\}=\pm1}{}' \begin{cases}\sum_{n_q,n_p}\left[U_{jn_q}U_{n_qj}^\dagger - \left(U_{jn_q}U_{n_qj}^\dagger\right)^2\right], & j = k,\\ \sum_{n_q,n_p}\left[-U_{jn_q}U_{n_qk}^\dagger U_{kn_p}U_{n_pj}^\dagger\right], & j \neq k,\end{cases} \tag{B.19}$$

where the matrices $U$ have implicit dependence on the charge configuration $\{q_i\}$.

## B.3 Change of Basis

In the previous section we assumed that the initial state had definite site occupation numbers. More generally the initial state can have definite occupation in a different single particle basis. For example, we consider in the main text the half-filled Fermi-sea, which corresponds to having a particle in each of the lowest single particle eigenstates of $\hat{H}_{\mathrm{FS}} = -\sum_{\langle jk\rangle} \hat{c}_j^\dagger \hat{c}_k$. To use the expressions from the previous sections we must perform a change of basis.

Let $\tilde{c}_j$ be the operators with respect to which the state $|\psi\rangle$ has definite occupation. These are related to real space basis operators through

$$\tilde{c}_j = \sum_k V_{jk}^\dagger \hat{c}_k, \tag{B.20}$$

where the columns of the matrix $V$ are the single particle eigenstates corresponding to the state $\tilde{c}_j|vac\rangle$.

We can then use all of the above expressions for the correlators by using the matrix $V$ to transform to the correct basis. For example, in the density average we use the replacement

$$U_{jn_q} \to \sum_l U_{jl} V_{ln_q}, \qquad U^\dagger_{n_q j} \to \sum_l V^\dagger_{n_q l} U^\dagger_{lj} \tag{B.21}$$

to get

$$\langle\psi|e^{i\hat{H}(q)t}\hat{c}_j^\dagger \hat{c}_j e^{-i\hat{H}(q)t}|\psi\rangle = \sum_{n_q,l,m} U_{jl} V_{ln_q} V^\dagger_{n_q m} U^\dagger_{mj}. \tag{B.22}$$

As another example, the determinant for the average magnetization we need to perform a similarity transformation $U \to V^\dagger U V$ to get

$$\langle\psi|e^{i\hat{H}(q)t} e^{-i\hat{H}_{jk}(q)t}|\psi\rangle = \det[V^\dagger U^\dagger U_{jk} V]_{n_i n_i}. \tag{B.23}$$

In a similar way all correlators can be computed using single-particle matrix products and determinants in the appropriate basis.

### Reference

1. Kovrizhin DL, Chalker JT (2010) Multiparticle interference in electronic Mach-Zehnder interferometers. Phys Rev B 81:155318. https://doi.org/10.1103/PhysRevB.81.155318

# Appendix C
# Krylov Subspace Decomposition

A major bottleneck for computing dynamics using exact diagonalization is the memory requirement for storing many-body wave functions. Although this can be drastically reduced if the model has conserved quantities, memory is still the limiting factor due to exponential growth of the Hilbert space dimension. A way around this for computing dynamical quantities is to use a smaller set of basis states to perform the time evolution via short time steps. An optimal basis of states for this method can be identified with the Krylov subspace generated by the Hamiltonian. In this Appendix we provide an outline of this method.

Our goal is to efficiently calculate the time evolution of a quantum state $|\Psi(t)\rangle = e^{-i\hat{H}t}|\Psi\rangle$ for which we use the Krylov subspace

$$\mathcal{K}_R = \text{span}\{|\Psi\rangle, \hat{H}|\Psi\rangle, \hat{H}^2|\Psi\rangle, \ldots, \hat{H}^{R-1}|\Psi\rangle\}, \tag{C.1}$$

where $\hat{H}$ is our Hamiltonian, $|\Psi\rangle$ is an initial state, and $R$ is the chosen number of states in Krylov subspace. The idea is that at short enough times the state $|\Psi(t)\rangle$ will be predominantly in this subspace as can be seen from a Taylor expansion of the unitary time evolution

$$|\Psi(t)\rangle = \sum_{n=0}^{\infty} \frac{(-i\hat{H}t)^n}{n!}|\Psi\rangle. \tag{C.2}$$

Given a basis for the Krylov space $\{|v_i\rangle\}$, the best approximation to the unitary time evolution is given by $e^{-i\hat{\mathcal{H}}t}$ with

$$\hat{\mathcal{H}} = \sum_{ij} |v_i\rangle\langle v_i|\hat{H}|v_j\rangle\langle v_j|. \tag{C.3}$$

Therefore, we need to diagonalize a matrix of dimension $R$ with matrix elements $\mathcal{H}_{ij} = \langle v_i|\hat{H}|v_j\rangle$. For a Hermitian operator the reduced Hamiltonian $\mathcal{H}$ takes a simpler tridiagonal form which can be efficiently diagonalized.



A. Smith, *Disorder-Free Localization*, Springer Theses,
https://doi.org/10.1007/978-3-030-20851-6

One of the main practical considerations is the accuracy of the Krylov subspace method and the orthogonality between Krylov basis vectors. Due to limited numerical precision, the computed Krylov eigenvectors will diverge from the true ones as more of them are included. A way around this is to orthogonalize each new vector to the previous set. However, the numerical errors that arise in this procedure make it also problematic, and eventually orthogonality will be lost. To get around these issues, after roughly 25 applications of matrix multiplications (the number is chosen empirically) we orthonormalize the entire set of vectors using efficient QR decomposition before proceeding further.

The accuracy of this approximation can be kept below a prescribed threshold only for a finite value of $t$, which is set by the size of the basis dimension $R$. To study time evolution on longer timescales we use a 'restarted evolution method', which computes the time evolution up to a certain time $\delta t$ and then repeats with the new starting state $|\Psi(t+\delta t)\rangle$ until the desired time is reached. To check the accuracy of the method we then perform the reverse time evolution and check the difference between the values on the forward and backward pass, which provides a good estimate of the deviation from the true value [1, 2].

The method described in this Appendix is limited by the memory required to store the Hamiltonian matrix $H$ and the Krylov basis states. Since the Hamiltonian is typically sparse and has $O(\alpha N)$ non-zero values, the memory requirement scales as $O((R+\alpha)N)$ compared with $O(N^2)$ for exact diagonalization. In our calculations we take $R = 50$, and $\delta t = 1.2$ and compute values for $dt = 0.2$, which we find gives acceptable errors of only 1–2 orders of magnitude above machine precision on the scale of the full time evolution. We note that the computational cost also scales linearly with the number of time steps.

## References

1. Sidje RB (1998) Expokit: a software package for computing matrix exponentials. ACM Trans Math Softw 24:130–156. www.expokit.org
2. Brenes M, Varma VK, Scardicchio A, Girotto I (2017) Massively parallel implementation and approaches to simulate quantum dynamics using Krylov subspace techniques. arXiv:1704.02770. https://arxiv.org/pdf/1704.02770.pdf

# Appendix D
# Kernel Polynomial Method

The Kernel Polynomial Method (KPM) of Ref. [1] is a efficient numerical technique for computing spectral quantities. In this thesis it is used to compute the single-particle density of states, where it allows us to access systems sizes much greater than possible with more direct methods. In 1D we also use the density of states obtained by the method to compute the energy resolved localization length, see Eq. (3.4).

The KPM expands a function with finite support into Chebyshev polynomials with modifications to the coefficients to both damp Gibbs oscillations and to increase the accuracy of the approximation. The benefit of the KPM is that calculations can be reduced to repeated multiplications of the Hamiltonian matrix, which is very efficient for sparse matrices. We will briefly describe the basic procedure, also see Ref. [1] for more details and examples.

## D.1 Chebyshev Expansion and Modified Moments

Chebyshev polynomials are defined on the interval $[-1, 1]$ and form an orthogonal basis with respect to inner products defined on this interval with a special weight function. For Chebyshev polynomials of the first kind $T_n(x)$ this weight function is $w(x) = (\pi\sqrt{1-x^2})^{-1}$, so that the inner product reads,

$$\langle f|g\rangle_1 = \int_{-1}^{1} \mathrm{d}x\, \frac{f(x)g(x)}{\pi\sqrt{1-x^2}}, \quad \langle T_n|T_m\rangle_1 = \frac{1+\delta_{n,0}}{2}\delta_{n,m}. \tag{D.1}$$

Chebyshev polynomials of the second kind $U_n(x)$ are defined with respect to the weight function $w(x) = \pi\sqrt{1-x^2}$, i.e.

$$\langle f|g\rangle_2 = \int_{-1}^{1} \mathrm{d}x\, \pi\sqrt{1-x^2}\, f(x)g(x), \quad \langle U_n|U_m\rangle_2 = \frac{\pi^2}{2}\delta_{n,m}. \tag{D.2}$$



A. Smith, *Disorder-Free Localization*, Springer Theses,
https://doi.org/10.1007/978-3-030-20851-6

These polynomials obey the useful recursion relations

$$\begin{aligned} T_{n+1}(x) &= 2xT_n(x) - T_{n-1}(x), \\ U_{n+1}(x) &= 2xU_n(x) - U_{n-1}(x), \end{aligned} \tag{D.3}$$

where $T_0(x) = 1$, $T_1(x) = x$ and $U_0(x) = 1$, $U_{-1}(x) = 0$. We will also use the explicit form $T_n(x) = \cos(n \arccos(x))$ for the polynomials of the first kind.

Given an orthogonal basis, together with the inner product, we can expand a function defined on the interval $[-1, 1]$ as

$$f(x) = \alpha_0 + 2\sum_{n=1}^{\infty} \alpha_n T_n(x), \tag{D.4}$$

where the moments, $\alpha_n$, are given by

$$\alpha_n = \langle f | T_n \rangle_1 = \int_{-1}^{1} \mathrm{d}x \, \frac{f(x)T_n(x)}{\pi\sqrt{1-x^2}}. \tag{D.5}$$

However, the numerical integration in Eq. (D.5) is problematic due to the square root appearing in the denominator. To get around this we instead define the functions

$$\phi_n(x) = \frac{T_n(x)}{\pi\sqrt{1-x^2}}, \tag{D.6}$$

which have the property $\langle \phi_n | \phi_m \rangle_2 = \langle T_n | T_m \rangle_1$, and thus we can instead write the expansion as

$$f(x) = \frac{1}{\pi\sqrt{1-x^2}} \left[ \mu_0 + 2\sum_{n=1}^{\infty} \mu_n T_n(x) \right], \tag{D.7}$$

where the moments $\mu_n$ are given by

$$\mu_n = \langle f | \phi_n \rangle_2 = \int_{-1}^{1} \mathrm{d}x \, f(x) T_n(x). \tag{D.8}$$

We now have to discretize the function argument and truncate the infinite sum. To make use of the properties of Chebyshev functions, we choose a set of $K$ points, $x_k = \cos(\pi(k + 1/2)/K)$, for $k = 0, \ldots, K-1$. With this choice the expansion takes the form

$$f(x_k) \approx \frac{1}{\pi\sqrt{1-x_k^2}} \left[ \mu_0 + 2\sum_{n=1}^{M-1} \mu_n \cos\left(\frac{\pi n(k+1/2)}{K}\right) \right], \tag{D.9}$$

where we truncated the sum, retaining the first $M$ terms of the expansion. Here we have used the explicit closed form for the Chebyshev polynomials, $T_n(x) = \cos(n \arccos(x))$.

### D.1.1 Rescaling

To use this expansion in terms of Chebyshev polynomials we must first rescale the energies and the Hamiltonian so that the bandwidth lies in the interval $[-1, 1]$. We thus define the rescaled Hamiltonian and energies

$$\tilde{H} = (H - b)/a, \qquad \tilde{E} = (E - b)/a, \tag{D.10}$$

where $a = (E_{\max} - E_{\min})/(2 - \epsilon)$ and $b = (E_{\max} + E_{\min})/2$. We include a small factor $\epsilon > 0$ to avoid stability problems near $\pm 1$. In practice we can use analytically obtained bounds on $E_{\max/\min}$ to avoid computing them explicitly. One could also compute $E_{\max/\min}$ for smaller system sizes, add a margin of error and use these for the bounds.

### D.1.2 Modified Moments: Gibbs Oscillations

Since we are expanding in periodic functions, the truncation of the sums leads to Gibbs oscillations. If we keep the first $M$ terms in the sum, then to remove these oscillations we introduce a Kernel of order $M$

$$K_M(x, y) = g_0\phi_0(x)\phi_0(y) + 2\sum_{m=1}^{M-1} g_m\phi_m(x)\phi_m(y), \tag{D.11}$$

which we use to define

$$f_{\mathrm{KPM}}(x) = \int_{-1}^{1} \mathrm{d}y\, \pi\sqrt{1 - y^2} K_M(x, y) f(y). \tag{D.12}$$

We can then determine the coefficients $g_m$ in the Kernel by demanding that $f_{\mathrm{KPM}}$ is as close as possible to the true function $f(x)$. Closeness can be defined in a number of different ways, each of which leads to different set of coefficients. In our calculations we use the Jackson Kernel[1] defined by coefficients

[1]The Jackson Kernel is derived by trying to impose uniform convergence on a restricted interval $[-1 + \epsilon, 1 - \epsilon]$ for any $\epsilon$, i.e., $\max_{x \in (-1+\epsilon, 1-\epsilon)} |f(x) - f_{\mathrm{KPM}}(x)| \to 0$ as $M \to \infty$, while also trying to optimize the rate at which $g_j \to 1$ as $M \to \infty$.

$$g_m = \frac{1}{M+1}\left[(M-m+1)\cos\left(\frac{\pi m}{M+1}\right) + \sin\left(\frac{\pi m}{M+1}\right)\cot\left(\frac{\pi}{M+1}\right)\right]. \tag{D.13}$$

See Ref. [1] for the derivation of these coefficients and discussions of other choices of kernel. Kernel coefficients are then used to modify the moments in our expansion, and we arrive to the expression

$$f(x_k) \approx \frac{1}{\pi\sqrt{1-x_k^2}}\left[g_0\mu_0 + 2\sum_{m=1}^{M-1} g_m\mu_m \cos\left(\frac{\pi m(k+1/2)}{K}\right)\right]. \tag{D.14}$$

## D.2 Calculation of the Moments

The moments that appear in our expansion are typically of the form $\langle\beta|AT_n(H)|\alpha\rangle$, where $H$ is the $N \times N$ Hamiltonian matrix, $A$ is a matrix representing an operator and $|\alpha\rangle$ and $|\beta\rangle$ are two states. We need to compute $|\alpha_n\rangle \equiv T_n(H)|\alpha\rangle$, which can be done using the recursion relation

$$|\alpha_{n+1}\rangle = 2H|\alpha_n\rangle - |\alpha_{n-1}\rangle, \tag{D.15}$$

with $|\alpha_0\rangle = |\alpha\rangle$ and $|\alpha_1\rangle = H|\alpha\rangle$. If $\beta = \alpha$ and $A = \mathrm{I}$ we can use a property of Chebyshev polynomials, specifically $2T_m(x)T_n(x) = T_{m+n}(x) + T_{m-n}(x)$, to get

$$\mu_{2n} = 2\langle\alpha_n|\alpha_n\rangle - \mu_0, \quad \mu_{2n+1} = \langle\alpha_{n+1}|\alpha_n\rangle - \mu_1, \tag{D.16}$$

which reduces the number of matrix operations by approximately half.

We also need to compute moments that involve a trace over states. The latter can be computed efficiently using

$$\mathrm{Tr}[AT_n(H)] \approx \frac{1}{R}\sum_{n=1}^{R-1}\langle r|AT_n(H)|r\rangle, \tag{D.17}$$

where $N$ is the size of the matrix $H$, and $R \ll N$ is the number of chosen random vectors $|r\rangle$, which are defined through random variables $\varepsilon_{ri}$

$$|r\rangle = \sum_{i=0}^{N-1}\varepsilon_{ri}|i\rangle, \tag{D.18}$$

where $|i\rangle$ are the basis vectors with an identity in the $i^{th}$ entry. The random variables must satisfy

$$\overline{\varepsilon_{ri}} = 0, \qquad \overline{\varepsilon_{ri}\varepsilon_{r'j}} = \delta_{rr'}\delta_{ij}, \tag{D.19}$$

that is, they are uncorrelated with zero mean, and have unit mean for their absolute value. In all of our calculations we take $R = 30$.

## D.3 Density of States

As an example of the application of the method, let us consider the calculation of the density of states, which is defined as

$$g(E) = \frac{1}{N}\sum_{k=0}^{N-1}\delta(E - E_k). \tag{D.20}$$

The coefficients of the Chebyshev expansion are then given by

$$\begin{aligned}\mu_n = \int_{-1}^{1} dE\, \rho(E)T_n(E) &= \frac{1}{N}\sum_{k=1}^{N-1} T_n(E_k)\\ &= \frac{1}{N}\sum_{k=1}^{N-1}\langle k|T_n(H)|k\rangle\\ &= \frac{1}{N}\mathrm{Tr}[T_n(H)],\end{aligned} \tag{D.21}$$

which we can compute using the statistical trace and the expectation values as explained above.

In the main text we use the following parameters for the figures:
Figure 3.6a $\leftarrow N = 10^6$, $M = 1500$, $K = 2M$, $R = 30$;
Figure 3.6b $\leftarrow N = 10^6$, $M = 7500$, $K = 3M$, $R = 30$;
Figure 3.10a $\leftarrow N = (10^3)^2$, $M = 1500$, $K = 2M$, $R = 30$;
Figure 3.12b $\leftarrow N = (10^3)^2$, $M = 2500$, $K = 2M$, $R = 30$;
where $N$ is the number of sites, $M$ is the number of moments included in the expansion, $K$ is the number of discretization points, and $R$ is the number of random states used in the statistical trace.

### Reference

1. Weiße A, Wellein G, Alvermann A, Fehske H (2006) The kernel polynomial method. Rev Mod Phys 78:275–306. https://doi.org/10.1103/RevModPhys.78.275

# Appendix E
# Transfer Matrix Method

The transfer matrix method is an iterative numerical technique for calculating localization lengths in disordered single-particle models. It uses a limiting result about random matrices that allows us to compute Lyapunov exponents for energy eigenstates in the thermodynamic limit. These Lyapunov exponents are related to the localization length and quantify the exponential decay of the localized wavefunctions.

The application of the transfer matrix approach in the calculations of the localization length proceeds by considering a system that is cut up into slices, with slices connected via transfer matrices [1]. From these we can extract the eigenvalues of a limiting matrix that gives the Lyapunov exponents for our system. For instance, consider a 1D chain with the Hamiltonian

$$\hat{H} = -J\sum_j (\hat{c}_j^\dagger \hat{c}_{j+1} + \text{H.c}) - \sum_j h_i \hat{c}_j^\dagger \hat{c}_j. \tag{E.1}$$

The action of this Hamiltonian on an eigenstate $|\psi\rangle = \sum_i \psi_i |i\rangle$, where $|i\rangle$ is the state localized on site $i$ gives the relation

$$E\psi_i = -J\psi_{i+1} - J\psi_{i-1} - h_i\psi_i, \tag{E.2}$$

where $E$ is the eigenvalue of the state $|\psi\rangle$. This equation can be written in a compact form by introducing a transfer matrix

$$\begin{pmatrix} \psi_{i+1} \\ \psi_i \end{pmatrix} = \begin{pmatrix} -\frac{1}{J}(E+h_i) & -1 \\ 1 & 0 \end{pmatrix} \begin{pmatrix} \psi_i \\ \psi_{i-1} \end{pmatrix}. \tag{E.3}$$

In higher dimensions these equations have to be modified slightly, in particular we get

$$E\psi_i = -J\psi_{i+1} - J\psi_{i-1} - H_{\text{perp}}\psi_i, \tag{E.4}$$


A. Smith, *Disorder-Free Localization*, Springer Theses,
https://doi.org/10.1007/978-3-030-20851-6

where $\psi_i$ is now a vector of the values of the wavefunction within the slice $i$, and $H_{\text{perp}}$ is the matrix representation of the Hamiltonian within the slice. The transfer matrix equation assumes the form

$$\begin{pmatrix} \psi_{i+1} \\ \psi_i \end{pmatrix} = \begin{pmatrix} -\frac{1}{J}(E + H_{\text{perp}}) & -1 \\ 1 & 0 \end{pmatrix} \begin{pmatrix} \psi_i \\ \psi_{i-1} \end{pmatrix}. \tag{E.5}$$

Given the transfer matrix, we can compute the product along a long chain of length $L$,

$$Q_L = \prod_{i=1}^{L} T_i, \tag{E.6}$$

where $T_i$ is the transfer matrix for slice $i$. Oseledec's theorem states that there exists a limiting matrix

$$\Gamma = \lim_{L\to\infty} (Q_L Q_L^\dagger)^{1/2L}, \tag{E.7}$$

with eigenvalues $\exp(\gamma_i)$, where $\gamma_i$ are the Lyapunov exponents of the matrix $Q_L$. The smallest Lyapunov exponent describes the slowest growth of the wavefunction and corresponds to the inverse of the localization length $\lambda$.

More intuitively, we can consider $Q_L$ as the transfer matrix between the extreme ends of the chain

$$\begin{pmatrix} \psi_{L+1} \\ \psi_L \end{pmatrix} = Q_L \begin{pmatrix} \psi_2 \\ \psi_1 \end{pmatrix}, \tag{E.8}$$

so the eigenvalues of $Q_L$ describe the growth of the wavefunction along the length of the system. We can then take the smallest eigenvalue, $q$, of the matrix $Q_L$, and compute the localization length via

$$\lambda = \frac{L}{\log(q)}. \tag{E.9}$$

The procedure of computing a matrix product of a large number of matrices is numerically unstable since the matrix elements diverge or vanish exponentially. We therefore must orthonormalize the product after a few steps. The numerical procedure is as follows: we iteratively construct the product matrix $Q$ by applying randomly generated transfer matrices $T$. After the number of applications exceeds a predefined limit, or the amplitude of the elements of the matrix exceed a threshold, we store the logarithm of the eigenvalues and orthonormalize the matrix $Q$, which can be done efficiently using QR-decomposition. We continue applying $T$, storing eigenvalues and orthonormalizing until we have reached the length of the chain $L$. See below a pseudo-code of the algorithm for computing the localization length $\lambda$:

```
Q ← Id
l ← 1
while l ≤ L do
    count ← 1
    while max(abs(Q)) ≤ theshold or count ≤ limit do
        Initialise random T for slice
        Q ← T Q
        l++
        count++
    end
    b ← eig(Q)
    c ← c + log(b)
    Q ← orthonormal(Q)
end
b ← eig(Q)
c ← c + log(b)
lambda ← max(L/c)
```

## Reference


1. Kramer B, MacKinnon A (1993) Localization: theory and experiment. Rep Prog Phys 56:1469–1564. https://doi.org/10.1088/0034-4885/56/12/001

Printed in the United States
By Bookmasters